PLANT SCIENCE RESEARCH AND PRACTICES

RAPESEED

CHEMICAL COMPOSITION, PRODUCTION AND HEALTH BENEFITS

PLANT SCIENCE RESEARCH AND PRACTICES

Additional books in this series can be found on Nova's website under the Series tab.

Additional e-books in this series can be found on Nova's website under the e-book tab.

PLANT SCIENCE RESEARCH AND PRACTICES

RAPESEED

CHEMICAL COMPOSITION, PRODUCTION AND HEALTH BENEFITS

MONICA WHITE
EDITOR

nova publishers
New York

Library of Congress Cataloging-in-Publication Data

Names: White, Monica, 1972- editor.
Title: Rapeseed : chemical composition, production and health benefits / editor: Monica White.
Other titles: Plant science research and practices.
Description: Hauppauge, New York : Nova Science Publishers, Inc., [2016] | Series: Plant science research and practices | Includes index.
Identifiers: LCCN 2015043138 | ISBN 9781634842273 (hardcover)
Subjects: LCSH: Rapeseed--Composition. | Rapeseed--Health aspects.
Classification: LCC SB299.R2 R25 2016 | DDC 633.8/53--dc23 LC record available at http://lccn.loc.gov/2015043138

Published by Nova Science Publishers, Inc. † New York

Contents

PREFACE

Rapeseed is the traditional name of a large winter or spring annual oilseed crop from the *Brassicaceae* family, and is related to mustard, cabbage, broccoli, among others. As one of the leading oilseed crops, rapeseed is primarily grown for its oil content. This book presents research on the chemical composition, production and health benefits of rapeseed. The first chapter reviews the literature on rapeseed/canola derived bioactive peptides with emphasis on strategic production and processing methods as well as antihypertensive, antitumor, hypocholesterolemic, and multifunctional properties. Chapter two summarizes recent advances on the extraction processes of rapeseed-derived protein and bioactive compounds and reviews the possibilities as well as the challenges that these ingredients face regarding skin care, food and nutraceutical applications. Chapter three examines physical properties of fat filling produced with oil industry by-products – sunflower and rapeseed lecithin in comparison with widely used emulsifier- soy lecithin. Chapter four provides a comprehensive summary about contribution of mass spectrometric methods to analysis of rapeseed, reflecting their irreplaceability to the field of "Food control." The final chapter studies antioxidant activity in transgenic canola plants grown *in vitro*.

Chapter 1 – Bioactive peptides (BAPs) have demonstrated potential for application as health-promoting agents against numerous human health and disease conditions, including cardiovascular disease, inflammation, and cancer. BAPs are encrypted in the primary structure of plant and animal proteins as inactive amino acid sequences but they can be released by fermentation, food processing, and enzyme-catalyzed proteolysis *in vitro* or in the digestive tract after human consumption. The feasibility of pharmacological application of these peptides depends on absorption and bioavailability in intact forms in target tissues, which in turn depends on structure of the peptides. Rapeseed proteins are considered to be attractive and promising sources of BAPs. Recently, a number of research works have focused on the investigation of different methods to produce active peptides from the rapeseed proteins. The importance of bioinformatic-driven approaches (often referred to as *in silico*) in the discovery of bioactive peptide encrypted in food proteins is growing. Classical empirical approaches are typically time consuming and expensive, and bioinformatics provide a cost-effective strategy through utilizing the information accrued in databases and enable the reduction of different steps of the empirical workflow. Thus far, published data concerning the discovery of BAPs encrypted in rapeseed proteins does not include utilization of *in silico* simulation. This is partly due to the limited knowledge on rapeseed protein-derived BAPs peptide sequences in the databases, but however, bioinformatics approach could be combined to classical approach to enhance the discovery and use of rapeseed protein-derived BAPs in functional food and health applications. Response surface methodology has demonstrated to be very useful in the optimization of proteolysis variables, such as time, temperature, pH, enzyme or bacterial strain when searching for a specific activity. This article reviews the literature on rapeseed/canola derived BAPs with emphasis on strategic production and processing methods as well as antihypertensive, antitumor, hypocholesterolemic, and multifunctional properties.

Chapter 2 – Rapeseed is the traditional name of a large winter or spring annual oilseed crop from the *Brassicaceae* family, and is related to mustard, cabbage, broccoli, among others. As one of the leading oilseed crops, rapeseed is primarily grown for its oil content. Production of rapeseed oil by cold or hot screw pressing solely or followed by solvent extraction, respectively, generates a significant co-stream in the form of press cake or meal. This co-stream is an excellent feedstock for nutritionally valuable components such as protein, and bioactive compounds such as phenolics and peptides, not only for livestock but also for human consumption. The mentioned bioactive compounds have been found to possess potential health benefits and skin-repairing properties and thus may play an important role not only in the cosmetics but also in the treatment or prevention of many diseases. Rapeseed protein, constituting 32-36% of the rapeseed press cake or meal, has a well-balanced amino acid composition and suitable technological properties for food use. However, there are still challenges that limit the use of rapeseed press cake for commercial profitable production of protein and bioactive compounds, particularly in relation to the organoleptic properties and levels of harmful or anti-nutritional factors such as glucosinolates and phytates. This chapter reviews the possibilities as well as the challenges that rapeseed press cake co-stream ingredients face regarding skin care, food and nutraceutical applications.

Chapter 3 – Vegetable lecithins are commercially obtained from oil-bearing seeds such as soybeans, sunflower and rapeseed. Soybean seeds are the main source of lecithin in confectionery industry in Serbia and elsewhere. On the other hand, production of sunflower and rapeseed oil makes economically sustainable production of sunflower and rapeseed lecithin and their use as emulsifiers in different confectionery products. This chapter examined physical properties of fat filling produced with oil industry by-products – sunflower and rapeseed lecithin in comparison with widely used emulsifier - soy lecithin. Rheological, textural and thermal properties were determined by rotational viscometry, texture analyser, and differential scanning calorimetry while particle size distribution was analysed using particle size analyser. The obtained results showed similar particle size distribution of fat filling produced with soy, sunflower and rapeseed lecithin within each applied milling time, indicating appropriate production. Regarding rheological properties, fat filling samples with sunflower lecithin had lower values of viscosity and textural parameters compared to samples with soy and rapeseed lecithin at all applied milling time. On the other hand, soy lecithin influenced higher values of mentioned parameters. Prolonging the milling time had no significant influence on termal parameters of fat filling produced with all three type of lecithin where samples with sunflower lecithin began to melt at lower temperatures in relation to samples with soy and rapeseed lecithin.

Chapter 4 – Rape plant is economically important oilseed crop, contributing up to 15% of oil production worldwide. European Union Directive 2009/28/EC mandatory determines for member States 10% sharing of energy from renewable sources by 2020, thus concentrating much effort in implementation of rapeseed fatty acid methyl esters as biodiesel, yielding to established standards in blends with petrodiesel, having benefits over diesel fuel such as decreasing in CO_2 and smoke density (European Standards EN 590). But rapeseed has beneficial nutritional properties, providing most important nutrition for human daily life: energy, essential fatty acids, phytosterols, and tocopherols. It has high protein content; over 200 flavonoids; and secondary metabolites showing antioxidant activity, prevention from cardiovascular diseases and cancer. The oil is 46% of dry weight of seed. Meal has 40% proteins, 35% fibers, 8.5% minerals and 1% lipids. Rape plant, thereby, is a fertilizing product of great importance for agriculture, too. The author's have chosen to emphasis on rapeseed as a food, calling attention to mass spectrometry for lipidomics, proteomics and low molecular weight analysis. The great applicability of mass spectrometry has been governed by high complexity of the foodstuff matrix, containing thousand components from biomacromolecules (12–21 kDa) to low molecular weight analytes in concentrations from macro-components (acylglycerols, > 94%) to traces (cytokinins, < pmol.g^{-1}). The analysis of agricultural pesticides is a challenging research task, too, due to lipophylic properties at trace levels, difficult their extraction. But, mass spectrometric methods have superior instrumental characteristics over a large scale of methods. These are: (*i*) instrumental flexibility and different ionization approaches, including off– and on-line hybrid operation; (*ii*) high–to–ultrahigh resolving power allowing analysis in multicomponent mixture. Particularly for rapeseed oil at resolution $m/\Delta m_{50\%} > 350\ 000$ and a mass accuracy < 1 ppm it has been achieved detail analysis of chemical content by electrospray ionization Fourier transform ion cyclotron resonance mass spectrometry without sample pretreatments; (*iii*) high accuracy and precision; (*iv*) low concentration limits of detection of *attomol* levels; (*v*) analytical information from macromolecules with masses higher than 50 kDa to low molecular weight analytes; and (*vi*) 3D imaging for assay from isolated living cell to whole organs and bodies. The chapter can be regarded as a comprehensive summary about contribution of mass spectrometric methods to analysis of rapeseed, reflecting their irreplaceability to the field of "Food control."

Chapter 5 – The research was aimed at studying the possible unintended biochemical peculiarities of the recently obtained herbicide-resistant canola (*Brassica napus* L.) plants which express both the *BAR* gene conferring resistance to glufosinate herbicides and synthetic *EPSPS* gene responsible for resistance to glyphosates. Both total radical scavenging activity and superoxide dismutase activity as a part of plant antioxidant system were investigated in the leaf extracts using DPPH (2,2-diphenyl-1-picrylhydrazyl radical) and nitroblue tetrazolium assays, respectively. The total soluble protein content was also evaluated. Significant decrease in the total radical scavenging activity was detected for the primary transgenic plants possessing two different genes for herbicide resistance. The total soluble protein content and superoxide dismutase activity were unaffected in these plants. The analyzed biochemical parameters in the leaf extracts from untransformed plants were similar to those of the herbicide-resistant plants of the first and the second generations grown *in vitro* on the media without herbicide supplementation. Addition of Basta (phosphinothricin) or Hurricane Forte (glyphosate) herbicide solutions to the culture media did not affect antioxidant activity of transgenic canola plants. Thus, T_2 generation homozygous canola plants resistant to both glufosinate and glyphosate herbicides had the total soluble protein content, total radical scavenging activity, and superoxide dismutase activity in their leaves similar to that of the untransformed control plants. Earlier the author's have produced several groups of transgenic canola lines bearing target genes of the different origins. There were the plants expressing either 1) both DesC desaturase (*Des*C) gene from cyanobacterium *Synechococcus vulcanus* and *EPSPS* gene or 2) human interferon α2b (*HuIFN*-α2b) gene or 3) *CYP11A1* gene for cytochrome $P450_{SCC}$ from bovine adrenal cortex mitochondria. Antioxidant activities in these plants have been compared to that of the newly obtained herbicide-resistant plants. The influence of different target genes on plant antioxidant activity and the importance of this parameter for both plants and human health are discussed.

In: Rapeseed
Editor: Monica White
ISBN: 978-1-63484-227-3

Chapter 1

Rapeseed Protein-Derived Bioactive Peptides: Production, Processing and Potential Health Benefits

Anne Pihlanto* and Sari Mäkinen
Natural Resources Institute Finland,
Bio-based Business and Industry, Jokioinen, Finland

Abstract

Bioactive peptides (BAPs) have demonstrated potential for application as health-promoting agents against numerous human health and disease conditions, including cardiovascular disease, inflammation, and cancer. BAPs are encrypted in the primary structure of plant and animal proteins as inactive amino acid sequences but they can be released by fermentation, food processing, and enzyme-catalyzed proteolysis *in vitro* or in the digestive tract after human consumption. The feasibility of pharmacological application of these peptides depends on absorption and bioavailability in intact forms in target tissues, which in turn depends on structure of the peptides. Rapeseed proteins are considered to be attractive and promising sources of BAPs. Recently, a number of research works have focused on the investigation of different methods to produce active peptides from the rapeseed proteins. The importance of

* Corresponding author: Natural Resources Institute Finland, Bio-based Business and Industry Myllytie 1 31600 Jokioinen Finland, phone +358 40 5133576, e-mail anne.pihlanto@luke.fi.

bioinformatic-driven approaches (often referred to as *in silico*) in the discovery of bioactive peptide encrypted in food proteins is growing. Classical empirical approaches are typically time consuming and expensive, and bioinformatics provide a cost-effective strategy through utilizing the information accrued in databases and enable the reduction of different steps of the empirical workflow. Thus far, published data concerning the discovery of BAPs encrypted in rapeseed proteins does not include utilization of *in silico* simulation. This is partly due to the limited knowledge on rapeseed protein-derived BAPs peptide sequences in the databases, but however, bioinformatics approach could be combined to classical approach to enhance the discovery and use of rapeseed protein-derived BAPs in functional food and health applications. Response surface methodology has demonstrated to be very useful in the optimization of proteolysis variables, such as time, temperature, pH, enzyme or bacterial strain when searching for a specific activity. This article reviews the literature on rapeseed/canola derived BAPs with emphasis on strategic production and processing methods as well as antihypertensive, antitumor, hypocholesterolemic, and multifunctional properties.

INTRODUCTION

During the past decades, a considerable interest in the area of functional foods and bioactive peptides has arised from both scientists and potential consumers. It is now recognized that some food proteins and peptides are able to elicit beneficial physiological effects and therefore have the potential to influence health and may help to reduce the risk of chronic disease. Peptides from food sources can be structurally similar to human endogenous regulatory peptides. Thus, it is reasonable that they can interact with the same receptors and play a role as modifiers on food intake, growth factors, immune regulators, or antimicrobials in the host organism (Pihlanto and Korhonen, 2003). These peptides can be found naturally in foods or released during hydrolysis of a precursor protein, being latent state as part of the sequence of that protein. As a result of proteolytic processes, that occur during *in vivo* digestion of proteins or food processing (fermentation, enzymatic digestion, autolysis), these peptides can be released showing certain activities. The type of bioactive peptides generated from a particular protein is dependent on two factors: (a) the primary sequence of the source protein and (b) the specificity of the enzyme(s) used to generate such peptides. The hydrolysis of plant proteins has led to the production of a variety of biologically active peptides, such as opioid, antihypertensive, antioxidative, immunomodulatory or antimicrobial peptides (Pihlanto and Korhonen, 2003; Guang and Philips, 2009). Bioactivity of peptides depend on the structure, however, the structure activity relationship is not yet fully understood for all biological activities described.

In the search for new peptides with biological activity, bioinformatics constitutes an important tool. Bioinformatics enables the prediction of protein structure-function relationships, the identification of putative protein sources of BAP and computer simulation of proteolytic processes. All this information can be extracted from the vast number of bioactive peptides that have already been isolated and identified. The structural motifs in active peptides serve as a source of information to be used in the search for new bioactive molecules. Databases, such as BIOPEP, can be utilized to determine the occurrence frequency of known BAPs and structure-function patterns in the primary sequence of food proteins and to simulate the proteolysis of dietary proteins based on the knowledge of the specific cleavage sites of certain enzymes. The resulting peptides from the simulated hydrolysis treatment can be matched with the BAPs with known bioactivities, and the remainder of the purportedly "inactive" peptides can be analyzed with quantitative structure activity relationship model (QSAR) to predict biological activities by physicochemical descriptors associated with known bioactivities. However, there are certain limitations with the *in silico* approaches. For example, it is not certain that *in silico* peptides can be reproduced empirically due to the complex nature of protease-substrate interactions. Also, the information included to databases is limited and often does not include constantly accumulating information on new bioactive peptides. Thus, an integrated approach utilizing *in silico* analysis in combination with experimental validation with hydrolysates and synthetic peptides is the most useful for mining and discovering of BAPs.

Rapeseed protein is suggested to be an attractive source of bioactive compounds. Recently, various different methods have been utilized to produce active peptides from rapeseed proteins. Most of the studies concern ACE inhibitory and antioxidant activity *in vitro* and antihypertensive effects in animal studies, but also other activities such as antithrombotic activity and cell growth effects have been demonstrated as bioactive characteristics of rapeseed proteins. This review has been focused on rapeseed-protein-derived bioactive peptides with antihypertensive and antioxidant activities, in addition to other less frequent activities.

Release and Identification of Bioactive Peptides

Bioactive peptides can be released by fermentation, food processing and enzyme-catalyzed proteolysis *in vitro* or in the digestive tract after human consumption (Hartmann and Meisel, 2007). Numerous animal and plant protein sources have been exploited as sources of bioactive peptides. Most widely studied protein sources are of animal origin, such as milk proteins and egg, but marine and plant protein sources have attracted increasing attention in recent years. Various plant protein sources have been utilized for bioactive peptide production, of which soy, pulses (lentils, peas, beans, chickpeas), cereals, canola and flaxseed are the typical sources. According to the current literature, enzymatic hydrolysis with single or multiple specific or nonspecific proteases has been the main process for producing BAPs. Number of studies concern production of Angiotensin I converting enzyme (ACE) inhibitory and antioxidative peptides from food proteins (for example Pedroche et al., 2002; Wu and Ding, 2002; Yust et al., 2003; Megías et al., 2006; Barbana and Boye, 2010; Udenigwe and Aluko, 2010; Udenigwe et al., 2012). Industrial food-grade proteinases such as Alcalase, Flavorzyme, and Protamex derived from microorganisms, as well as enzymes from plant (e.g., Papain) and animal sources (e.g., Pepsin and trypsin), have been widely used in producing ACE inhibitory and antioxidative peptides. During the enzymatic production of bioactive peptides several factors affect the resulting peptide composition, and their structural and physicochemical characteristics. These factors include the specificity of the enzyme used for hydrolysis, hydrolysis time, enzyme to substrate ratio, and pretreatment of the protein prior to hydrolysis. For example, thermal treatment of proteins can enhance enzymatic hydrolysis (Inouye et al., 2009). On the other hand, a number of studies have shown that ACE inhibitory peptides are liberated during fermentation. In particular, milk products fermented with lactic acid bacteria strains have been found to possess antihypertensive and antioxidative capacity (Virtanen et al., 2007; Pihlanto et al., 2010). Furthermore, several antioxidative and ACE inhibitory peptide sequences have also been identified from fermented milk (Hernández-Ledesma et al., 2005). Fermentation has also been shown to release ACE inhibitory and antioxidant peptides from plant proteins, such as soybean. For example, the traditional fermented soybean products natto, tempeh and douche contain antioxidative and ACE inhibitory peptides released by fungal proteases. Enzymatic hydrolysis has been shown to produce higher potent bioactivities than fermentation especially from plant proteins.

In the case of rapeseed, most of the studies concern the utilization of various proteolytic enzymes to liberate ACE inhibitory peptides (Marczak et al., 2003; Mäkinen et al., 2012; He et al., 2013c; He et al., 2013a; He et al., 2013b; Alashi et al., 2014). However, enzymatic hydrolysis of rapeseed proteins has produced also antioxidative, renin inhibitory, antithrombotic and hypocholesterolemic activities. Also, fermentation with lactic acid bacteria has been shown to release *in vitro* ACE inhibitory peptides from rapeseed (Pihlanto et al., 2012). The production methods and bioactive properties of rapeseed protein-derived digests are summarized in Table 1. Enzymatic hydrolysates of rapeseed meal and protein extracts have shown high ACE inhibitory potencies *in vitro*; the most active hydrolysates have been produced by Alcalase with 50% ACE inhibition concentrations between 15 and 20 μg protein/mL (Wu et al., 2008; Mäkinen et al., 2012; He et al., 2013a). Also, Thermolysin and consecutive hydrolysis with Alcalase and ENZECO Alkaline Protease L-FG have produced high ACE inhibitory capacity (He et al., 2013a; Wu et al., 2008). Rapeseed protein hydrolysates have also been shown to possess inhibitory capacity against renin, the other key enzyme in the regulation of blood pressure through the RAS system. In particular, the rapeseed protein Alcalase hydrolysate and Pancreatin hydrolysate have been reported to possess dual inhibitory capacity against ACE and renin, both key factors in the regulation of blood pressure (He et al., 2013a; Alashi et al., 2014)

Antioxidant effects of rapeseed protein hydrolysates and peptides have been observed with various different *in vitro* methods, either related to scavenging free radicals or inhibition of lipid peroxidation and formation of malondialdehyde (MDA). Most of the studies concern the utilization of commercial proteases for the production of antioxidant peptides, in particular, Alcalase has produced highly effective antioxidant capacity from rapeseed protein. Xue et al., (2009a) investigated the digestion of insoluble rapeseed meal protein with Alcalase and Flavourzyme to produce peptides with antioxidant properties and observed dose-dependent reducing antioxidant activities and hydroxyl radical scavenging abilities. The hydrolysate also inhibited malonyldialdehyde (MDA) generation by 50% in blood serum at 150 mg/ml. The hydrolysate was also found to inhibit the autohemolysis of rat red blood cells and MDA formation in a rat liver tissue homogenate (Xue et al., 2009b). The rapeseed protein-derived Alcalase hydrolysate has also demonstrated radical scavenging activity against DPPH radical and inhibition capacity against lipid peroxidation in a liposome system with comparable effects to those of ascorbic acid at 5 mg/mL (Zhang et al., 2008; Zhang et al.,

2009; Mäkinen et al., 2012). In addition, two other studies have reported high antioxidant activities of rapeseed protein hydrolysates against free radicals, and in particular Alcalase hydrolysate showed high potential with the oxygen radical absorbance capacity (ORAC) method (He et al., 2013a; He et al., 2013b).

Table 1. Production methods of rapeseed protein-derived BAPs

Treatment: enzyme/microbial strain	*Outcome*	*Reference*
Pepsin, Trypsin, Alcalase, Subtilisin, Thermolysin, Pancreatin, Chymotrypsin *Bacillus subtilis, Lactobacillus helveticus*	High ACE inhibition *in vitro*	Marczak et al., 2003; Wu et al.,2008; Mäkinen et al., 2012; Pihlanto et al., 2012; He et al., 2013a;Yu et al., 2013
Alcalase, Thermolysin, Pepsin, Pepsin+Pancreatin, Chymotrypsin, Flavourzyme, Proteinase K	Renin inhibition *in vitro*	He et al., 2013a; Alashi et al., 2014
Subtilisin, Pepsin, Alcalase, Thermolysin, Chymotrypsin, Pancreatin, Pepsin+Pancreatin, Flavourzyme, Proteinase K	Hypotensive effects *in vivo* on spontaneously hypertensive rats (SHR)	Marczak et al., 2003; He et al., 2013a; He et al., 2013b Alashi et al., 2014,
Pepsin, Pepsin+Pancreatin, Trypsin, Alcalase, Subtilisin, Thermolysin, Proteinase K, Flavourzyme *B. subtilis, L.helveticus*	Antioxidant activities: Radical scavenging activity, inhibition of lipid peroxidation in liposome micelles, inhibition of linoileic acid oxidation, inhibition of autohemolysis of rat red blood cells, inhibition of themalonyldialdehyde generation in blood serum	Zhang et al., 2008; Xue et al., 2009a;Xue et al.,2008b;Mäkinen et al., 2012; Pihlanto et al., 2012; Yu et al., 2013; He et al., 2013a; He et al., 2013b
Pepsin, Pepsin/Pancreatin, non digested rapeseed protein	Hypocholesterolemic properties: bile acid-binding *in vitro*	Yoshie-Stark et al., 2008
Alcalase	Antithrombotic activity: inhibition of thrombin-catalyzed coagulation of fibrinogen *in vitro*	Zhang et al., 2008
Alcalase+Flavourzyme	Antitumor and antioxidative effects	Xue et al., 2009b

Results have demonstrated a notable inverse relationship between radical scavenging activity (ORAC) and peptide molecular weight; the ORAC values were highest for the < 1 kDa peptides but decreased as peptide size increased to 5 - 10 kDa indicating that large-size peptides may not interact efficiently with the free radical (He et al., 2013a). On the other hand, Pihlanto and co-workers (2012) investigated the inhibition of lipid peroxidation of fermented rapeseed meal and different molecular weight fractions using a liposome model. Fermentation with *L. helveticus* and *B. subtilis* strains produced the capacity to inhibit lipid oxidation and fractionation with size exclusion chromatography separated the inhibition activity against lipid peroxidation in the 1 200 – 1 300 Da fraction. This indicates that the peptides with lipid peroxidation inhibitory properties are relatively higher in molecular mass than peptides with efficient radical scavenging properties. The difference could be due to the ability of larger peptides to bind into the liposome micelle structure.

Altogether, due to the various different methods used for antioxidant determination, comparison of the reported results is not straightforward, but on the other hand, results show that rapeseed is a diverse source of different types of antioxidant peptides as the hydrolysates and fractions have shown the capacity to prevent degenerative oxidative damage through different mechanisms. Also, the published data demonstrate that rapeseed hydrolysates have potentially high antioxidant potency as compared to other hydrolysates found in literature. For example, soy protein hydrolysates with Alcalase yielded ORAC values between 23 and 83.8 μM TE/g protein (Zhang et al., 2010) while amaranth protein hydrolysate produced by PP had a value of about 75 μM TE/g protein (Orsini Delgado et al., 2011), which are substantially lower than the values obtained with rapeseed hydrolysates. However, the nonprotein compounds, such as phenols and pigments need to be noted when evaluating the rapeseed samples. Rapeseed meal is known to be rich in phenolic compounds, most importantly sinapic acid derivatives (Vuorela et al., 2004). In the published data Alcalase has been consistently shown to be an excellent protease for releasing both ACE inhibitory and antioxidant peptides from rapeseed proteins.

Identified ACE Inhibitory and Antioxidant Peptide Sequences

Several ACE inhibitory peptides have been isolated and identified from the enzymatic hydrolysates of rapeseed protein. All of the identified peptide sequences have been produced by either with Subtilisin or Alcalase. However, only few rapeseed protein-derived antioxidative peptides have been identified thus far.

Table 2. Identified BAPs from rapeseed protein-derived hydrolysates

Peptide sequence	*ACE inhibitory IC50valueµM*	*Renin inhibitory IC50value mM*	*Antioxidant activity*	*Reference*
Arg-Ile-Tyr	28			Marczak et al., 2003
Ile-Tyr	37			Marczak et al., 2003
Val-Trp	1.6			Marczak et al., 2003
Val-Trp-Ile-Ser	30 mM			Marczak et al., 2003
Val-Ser-Val	0.15			Wu et al., 2008
Phe-Leu	1.33			Wu et al., 2008
Leu-Tyr	110	1.87		He et al., 2013b
Thr-Phe	810	3.1		He et al., 2013b
Arg-Ala-Leu-Pro	970	0.97		He et al., 2013b
Pro-Phe-Asp-Ser-Tyr-Phe-Val-Cys	0.19 mg/ml		radical scavenging	Yu et al., 2013
Gly-His-Ser	0.52 mg/ml	0.32 mg/ml		He et al., 2013c
Pro-Ala-Gly-Pro-Phe			radical scavenging	Zhang et al., 2009

The identified rapeseed protein-derived BAP sequences are summarized in Table 2. Four ACE inhibitory peptides have been isolated from the rapeseed Subtilisin digest: Arg-Ile-Tyr, Ile-Tyr, Val-Trp and Val-Trp-Ile-Ser (Marczak et al., 2003) of which Ile-Tyr and Arg-Ile-Tyr can be found in the primary structure of napin and Val-Trp and Val-Trp-Ile-Ser exist in the primary structure of cruciferin and ribosomal protein, respectively. ACE inhibitory peptides Val-Ser-Val, Phe-Leu, Leu-Tyr, Thr-Phe and Arg-Ala-Leu-Pro have been identified from Alcalase hydrolysate of rapeseed meal (He et al., 2013b) and Gly-His-Ser from the 3 kDa membrane ultrafiltration permeate of a Pepsin+Pancreatin rapeseed protein digest (He et al., 2013c). These peptides are located in the primary structure of rapeseed napin and cruciferin proteins (Wu et al., 2008). Interestingly, peptides Leu-Tyr, Thr-Phe, Arg-Ala-Leu-Pro and Gly-His-Ser have been shown to possess dual inhibition potential against ACE and renin activity (Table 2).

Processing Methods

A challenge frequently faced in food protein-derived peptide research is how to process the protein hydrolysates further in order to obtain high-yield peptide products with enhanced bioactivity. Careful choice of suitable enzymes and digestion conditions such as optimal temperature, degree of hydrolysis and enzyme-substrate ratio, as well as the control of hydrolysis time, is crucial for obtaining protein hydrolysates with desirable functional and bioactive properties. Hydrolysis can be performed by conventional batch hydrolysis or by continuous hydrolysis using ultrafiltration membranes. The traditional batch method has several disadvantages, such as the relatively high cost of the enzymes and their inefficiency compared to a continuous process, as noted in numerous studies (Mannheim and Cheryan, 1990; Wang et al., 2010). The hydrolysis process is feasible to scale-up production of peptides from laboratory scale to pilot and industrial plant scales with conserved peptide profiles and bioactivity of the resulting products (Chiang et al., 1995).
After protein hydrolysis, the resulting bioactive peptides can be concentrated based on physicochemical and structural characteristics of the peptides of interest, to enhance the bioactivity. The peptide properties that are often focused on include size, net charge and hydrophobicity, depending on the targeted bioactivities. The crude protein hydrolysate can be further processed by being passed through ultrafiltration membranes and size-exclusion chromatography in order to obtain a more uniform product with the desired range of molecular mass, especially to concentrate low molecular-weight peptides that are more resistant to further *in vivo* proteolytic digestion and more potential ACE inhibitors. Ultrafiltration membrane reactors have been shown to improve the efficiency of enzyme-catalysed bioconversion, to increase product yields, and to be easily scaled up. Furthermore, ultrafiltration membrane reactors yield a consistently uniform product with desired molecular mass characteristics (Mannheim and Cheryan, 1990; Guérard, 2007). Low molecular mass cut-off membranes are useful for concentrating bioactive peptides from the higher molecular mass components remaining, including undigested polypeptide chains and enzymes. In the case of rapeseed protein hydrolysates, ultrafiltration with 3 kDa membrane has been utilized to concentrate ACE inhibitory peptides into permeate (Mäkinen et al., 2012; Alashi et al., 2014) and moreover, to produce peptide concentrates showing hypotensive effects in SHR (Alashi et al., 2014). Also, our unpublished results have demonstrated hypotensive effects for rapeseed protein hydrolysate

concentrated with ultrafiltration with 3 kDa membrane in Goldblatt rat model of hypertension.

In addition to ultrafiltration techniques, reverse-phase chromatography on a hydrophobic column matrix can be used to fractionate peptides according to hydrophobicity, and ion exchange chromatography with selective columns can be utilized to produce peptide fractions of a particular net charge (Pownall et al., 2010; Pownall et al., 2011). These approaches are very advantageous especially when the peptide properties that inhibit the molecular disease targets are known. For example, antioxidant activities of peptide fractions have been demonstrated to correlate negatively with cationic property (Pownall et al., 2011). A novel membrane technology known as electrodialysis-ultrafiltration is useful for separating cationic, anionic and neutral peptides with defined molecular size on a large scale (Firdaous et al., 2009). Recently, this method has been utilized to concentrate and separate low molecular-weight bioactive peptides with a net positive or negative charge (Doyen et al., 2011a; Doyen et al., 2011b). Absorbent materials, such as activated carbon, are applicable for enriching particular amino acids in food protein hydrolysates (Udenigwe and Aluko, 2010). This processing method can result in sufficient yield if the amino acid residues of the target are prevalent in the peptides of interest. In addition, there are extensive bioassay-guided purification methods for the isolation of pure peptides for further analysis, especially for structure-function investigations. However, low peptide yield is still a limiting factor of the feasibility of using food protein-derived bioactive peptides. Therefore, it would be relevant to develop large-scale applicable food-grade processing methods for the production of a high yield of highly active peptide fractions. Understanding the unique structural characteristics of peptides with targeted bioactivity and exploitation of these characteristics in the concentration of the particular peptides is a crucial requirement for this approach.

HEALTH BENEFITS

Methods for Assaying Health Effects of BAPs *In Vitro* and *In Vivo*

The search for *in vitro* ACE inhibitors is the most common strategy followed in the selection of potential antihypertensive peptides derived from food proteins. ACE is a constituent enzyme of the Renin-Angiotensin-Aldosterone System (RAAS), which is a crucial regulator in human physiology. It controls blood pressure, fluid and electrolyte balance and affects the heart, vasculature and kidney (Lavoie and Sigmund, 2003). ACE inhibitory activity *in vitro* is generally evaluated by measuring the effect of potential inhibitors on the conversion of an appropriate substrate by ACE. There are several methods, and those based on spectrophotometric (Cushman and Cheung, 1971; Vermeirssen et al., 2002; Li et al., 2005) and high-performance liquid chromatography (HPLC) assays are the most commonly utilized (Doig and Smiley, 1993; Hyun and Shin, 2000; Shalaby et al., 2006). The most commonly utilized substrate for ACE is Hippuryl-His-Leu (HHL), and the methods measure the catalytic activity of ACE to produce hippuric acid from the substrate (Cushman and Cheung, 1971; Doig and Smiley, 1993; Li et al., 2005; Shalaby et al., 2006). In the widely used spectrophotometric method of Cushman and Cheung (1971) the hippuric acid released is measured after its extraction with ethyl acetate. The hippuric acid can also be measured by HPLC assays avoiding the extraction step (Vermeirssen et al., 2002). However, 2-furanacryloyl-phenylalanylglycyl-glycine (FAPGG) can also be used as substrate for ACE (Holmquist et al., 1979). This method has also been applied in microtiter plates (Vermeirssen et al., 2002). Moreover, substrates such as o-aminobenzoylglycyl-p-nitrophenylalanylproline are designed for fluorometric assays (Carmel and Yaron, 1978; Sentandreu and Toldrá, 2006).

On the other hand, another strategy for the selection of potential food-derived antihypertensive peptides is to search for *in vitro* renin inhibitory activity. Renin inhibition could provide a more complete blockade of the RAS, since renin catalyzes the rate-determining step in RAS (Fitzgerald et al., 2011). Recent studies have reported renin inhibition capacity, for example from rapeseed protein-derived hydrolysates and peptides (He et al., 2013a; He et al., 2013b; Alashi et al., 2014). The *in vitro* renin inhibitory capacities have been measured with a fluorometric method using a human recombinant renin inhibitor screening assay kit (Cayman Chemical, Ann Arbor, MI, USA). Interestingly, several rapeseed peptides have shown dual inhibition capacity both for ACE and renin (He et al., 2013b).

In vitro cultured cell model systems are used for screening of compounds for their bioavailability, metabolism and bioactivity to obtain valuable information before expensive and time-consuming *in vivo* studies. Endothelial cell models are commonly utilized, particularly in angiogenesis research. Endothelial dysfunction initiates atherosclerosis and the dysfunction is characterized by alteration in nitric oxide and endothelin-1 homeostasis (Verma et al., 2003).

There are several assays available to analyze the effects of bioactive compounds, such as bioactive peptides, on the endothelial function *in vitro*. For example, expression and activation of endothelial nitric oxide synthase – the enzyme responsible for the release of the vasodilator compound nitric oxide – can be analyzed in endothelial cells *in vitro* to elucidate the effects of the compounds of interest on the potential release of nitric oxide in the vascular endothelium.

Specific assays have not yet been developed or standardized to measure the antioxidative capacity of peptides or peptide mixtures. Therefore, assays that are commonly used for measuring antioxidative capacity of non-peptidic antioxidants have been used in the literature to measure the antioxidative capacity of peptides as well. Due to the complexity of oxidative processes occurring in food or biological systems as well as the different antioxidative mechanisms by which various compounds may act, finding one method that can characterize the overall antioxidative potential of food is not an easy task. Nevertheless, methods such as the Trolox equivalent antioxidant capacity (TEAC) assay, oxygen radical absorbance capacity (ORAC) assay, and the total radical-trapping antioxidant parameter (TRAP) assay have been widely reported in the literature for measuring antioxidative capacity of food and biological samples (Cao and Prior, 1998; Re et al., 1999). Commonly used assays include measuring the inhibition of lipid peroxidation in a linoleic acid model system and the capacity to scavenge the 2,2-Azino-bis (3ethylbenzo-thiaxoline-6-sulphonic acid)/2,2-Diphenyl-1-picrylhydrazyl (ABTS/DPPH) radicals.

Use of cell culture models for antioxidant research is particularly important since the studies to date have demonstrated that the mechanism of the action of antioxidants in human health promotion go beyond the antioxidant activity of scavenging free radicals (Liu and Finley, 2005). During experiments, intracellular oxidation of cells can be induced by using a peroxy radical generator or by using hydrogen peroxide (H_2O_2) (Elisia and Kitts, 2008). The 20,70-dichlorofluorescein diacetate (DCFH-DA) probe can be used to measure the extent of intracellular radical formation with and without added antioxidative compound in order to assess the cellular antioxidant activity (CAA) (Wolfe and Liu, 2007).

Table 3. Health enhancing properties *in vivo* of rapeseed protein-derived hydrolysates and peptides

Source material	***Treatment***	***Sample***	***Outcome***		***Reference***
Rapeseed meal	Subtilisin	Peptide Arg-Ile-Tyr (rapakinin)	Suppressed food intake and increased CCK release in SHR Vasorelaxing activity inmesenteric artery isolated from SHR		Marczak et al., 2006
Rapeseed meal	Alcalase+Flavourzyme	Protein hydrolysate	Antitumor and antioxidant effects in vivoin a murine model		Xue et al., 2009b
			Dose (mg prot./kg) in SHR, oral administration	Change in SBPΔ mmHgin comparision to control	
Rapeseed proteinisolate	Pepsin Subtilisin	Protein hydrolysate Protein hydrolysate Val-Trp Val-Trp-Ile-Ser Ile-Tyr Arg-Ile-Tyr	500 500 7.5 12.5 7.5 7.5	-7 after 4h -16 after 4h -11 after 2h -13 after 2h -10 after 2h -11 after 4h	Marczak et al., 2003
	Alcalase	Thr-Phe Leu-Tyr Arg-Ala-Leu-Pro	30 30 30	-12 after 4h -27 after 2h -17 after 2h	He et al., 2013b

Table 3. (Continued)

Source material	*Treatment*	*Sample*	*Outcome*		*Reference*
	Alcalase Proteinase K Pepsin+Pancreatin Thermolysin Flavourzyme	Proteinhydrolysate Protein hydrolysate Protein hydrolysate Protein hydrolysate Protein hydrolysate Rapeseed protein isolate	100 100 100 100 100 100	-24 after 8h -5 after 8h -21 after 24h -9 after 8h -17 after 6h -11 after 6h	He et al., 2013a
	Pepsin+Pancreatin Pepsin+Pancreatin	Gly-His-Ser 3kDa permeate of protein hydrolysate	30 100	-18 after 6h -21 after 4h	He et al., 2013c
Canola protein	Alcalase Chymotrypsin	Protein hydrolysate Protein hydrolysate	200	-35 after 4h -16 after 6h	Alashi et al., 2014
isolate	Pepsin Trypsin Pancreatin	Protein hydrolysate Protein hydrolysate Protein hydrolysate		-24 after 4h no change -15 after 6h	

The antihypertensive effects can be assessed by *in vivo* experiments using spontaneously hypertensive rats (SHR) that constitute an accepted animal model to study human essential hypertension (Erdmann et al., 2008). This animal model has been widely used to study the hypotensive effects of ACE inhibitory milk protein-derived peptides in both a short-term and long-term manner (Fitzgerald et al., 2004; Erdmann et al., 2008), and the current literature also includes studies concerning the *in vivo* antihypertensive effects of plant protein-derived peptides in SHR, e.g., pea (Vermeirssen et al., 2005), canola (Alashi et al., 2014) and flaxseed (Udenigwe et al., 2012). The published data concerning the *in vivo* measurement of the antioxidant effects of food protein-derived bioactive peptides is quite limited; however, biomarkers of lipid and protein peroxides as well as DNA damage can be assessed to monitor changes in oxidative stress *in vivo*.

Health Effects *In Vivo* of Rapeseed Protein Derived BAPs

As the number of *in vitro* studies has shown, rapeseed proteins comprise an attractive source of BAPs with potential antihypertensive properties. Recently, antihypertensive effects of rapeseed protein-derived BAPs have been shown also in animals. Also, other health enhancing properties such as antitumor and antioxidant effects have been reported for rapeseed protein-derived peptides in animals. The health enhancing properties demonstrated *in vivo* for rapeseed and canola protein-derived hydrolysates and peptides are summarized in Table 3.

Hypotensive Effects of Rapeseed Protein-Derived Peptides

Quite many rapeseed hydrolysates, fractions and peptides have recently been reported to induce antihypertensive effects on spontaneously hypertensive rats (SHR) (Table 2). Subtilisin digest of rapeseed protein has shown a dose-dependent antihypertensive effect after oral administration to SHR with a significant decrease in blood pressure even at a single dose of 0.15 g/kg body weight (bw). Rapeseed protein-derived Alcalase and Pepsin hydrolysates have also shown appreciable antihypertensive effects, with Alcalase hydrolysate producing the greatest (−34 mm Hg) and fastest (4 h) decrease in systolic blood pressure (SBP). Permeate of rapeseed protein Pepsin+Pancreatin digest (< 3 000 Da) has induced a corresponding hypotensive effect (−21 mmHg) after oral dosage on SHRs. In comparison, the non-hydrolyzed canola protein isolate has shown a significant and more prolonged (24 h) SBP-reducing effect, which is attributable to the extensive protein hydrolysis in the gastrointestinal tract (Alashi et al., 2014). In contrast, trypsin hydrolysate has been reported to be ineffective at reducing hypertension in SHR after oral administration (Marczak et al., 2003). The current literature also includes information on antihypertensive effects for rapeseed protein-derived pure peptides. Marczak and colleagues (2003) measured the antihypertensive activities of Val-Trp, Val-Trp-Ile-Ser, Ile-Tyr and Arg-Ile-Tyr following oral administration to SHR. All peptides displayed a dose-dependent antihypertensive effect. Hypotensive activity of the peptides was compared between young (19 - 20 weeks) and old (28 - 30 weeks) SHRs. Usually the hypotensive effect of ACE inhibitors in old SHR is lower than in young SHR, but Arg-Ile-Tyr (rapakinin) showed a similar effect both in old and young rats. Thus, it was suggested that another mechanism besides ACE inhibition may be involved in the hypotensive effect of rapakinin. Subsequently, the hypotensive effect of rapakinin was found to be mediated mainly by the prostaglandin-cholecystokinin-dependent vasorelaxation (Yamada et al., 2010). Recently, the antihypertensive effects of the peptides Leu-Tyr, Thr-Phe and Arg-Ala-Leu-Pro from Alcalase digest of rapeseed protein were tested on SHR (He et al., 2013b). Single oral administration to SHR showed Leu-Tyr and Arg-Ala-Leu-Pro to be the more effective hypotensive agents with a maximum blood pressure reduction of -26 and -16 mmHg, respectively, when compared to Thr-Phe (-12 mmHg). It was suggested that the higher number of hydrophobic amino acid residues of Leu-Tyr and Arg-Ala-Leu-Pro contributed to their higher *in vitro* and *in vivo* activities when compared to Thr-Phe (He et al., 2013b). Oral administration of Gly-His-Ser to SHR showed a maximum blood pressure reduction of −17 mmHg after 6 h. The *in vitro* ACE inhibitory effects of Gly-His-Ser were significantly weaker in comparison to the previously mentioned peptides and the hypotensive effects of Gly-His-Ser was suggested to be controlled mainly through renin inhibition (He et al., 2013b). The administration doses of single, pure rapeseed peptides reviewed above varied between 7.5 and 30 mg/kg bw, and the crude hydrolysates and fractions were administered in higher doses varying between 100 and 500 mg/kg bw.

Other Bioactivities

In addition to antioxidant and antihypertensive properties, some other potential health enhancing properties have also been reported for rapeseed protein-derived peptides (Table 2). Rapakinin (Arg-Ile-Tyr) derived from rapeseed napin, has been reported to possess food intake-suppressing properties, in addition to the hypotensive properties. Rapakinin dose-dependently decreased food intake and gastric emptying after oral administration at a dose of 150 mg/kg in mice (Marczak et al., 2006). Rapeseed protein has also been linked to preventing insulin resistance. Mariotti and others (2008) demonstrated that substituting rapeseed protein for milk protein prevented the early onset of insulin resistance, and the effects corresponded to those achieved by manipulating dietary fat and carbohydrates in a rat model.

Hypercholesterolemia is considered to be a major cause of heart diseases and atherosclerosis (World Health Organization, 2011). Healthy diet, exercise and consumption of bile acid sequestrants, which are better known as hypolipidemic agents, are important in the prevention of hyperlipidemia (Anderson and Siesel, 1990). Bile acid-binding promotes the consumption of cholesterol in the liver, which therefore reduces the level of cholesterol in the blood (Anderson and Siesel, 1990). The *in vitro* hypocholesterolemic properties of rapeseed proteins have been measured by determining the bile acid-binding capacity. Yoshie-Stark et al., (2008) demonstrated the bile salt-binding effects for rapeseed protein isolates and their hydrolysates, which were obtained by Pepsin and Pepsin/Pancreatin digestion. However, the hydrolysis treatments did not affect the bile acid-binding capacities indicating that some polypeptides present already in the rapeseed protein isolates are able to bind the ligands (Yoshie-Stark et al., 2008). Bile acid-binding properties of some other plant proteins and hydrolysates have also been reported with corresponding observations that hydrolysis does not necessarily enhance the bile acid-binding capacity (Yoshie-Stark and Waesche, 2004; Ma and Xiong, 2009).

Thrombosis is an anomaly in blood coagulation and leads to severe symptoms, heart injury among them. The blood-clotting process is related to the interaction between thrombin and fibrinogen to form the fibrin clot. Antithrombotics can reduce platelet aggregation and enhance fibrinolysis and thus, reduce the risk of thrombosis (Scheraga, 2004; Erdmann et al., 2008). Bioactive peptides have been demonstrated to possess antithrombotic activity, for example milk κ-casein-derived peptides and egg white protein hydrolysate (reviewed by Silva and Malcata, 2005). Antithrombotic activity has also been demonstrated for rapeseed peptides *in vitro* (Zhang et al., 2008). The rapeseed peptide fractions produced with Alcalase hydrolysis had significant inhibitory activities on the thrombin-catalyzed coagulation of fibrinogen. The maximum inhibitory effect was 90% at peptide concentrations between 30 and 50 mg/mL. The observed values were superior to those reported for egg white hydrolysate at the same concentration, but weaker than the antithrombotic activity of heparin, which is commonly used as an antithrombotic drug (Zhang et al., 2008).

Delivery and Bioavailability

Bioavailability is a major issue when establishing correspondence between *in vitro* and *in vivo* activities of bioactive peptides. The capacity to reach target organ in an active conformation determines the physiological effect of bioactive peptides. Various processes take place after oral administration of a bioactive peptide and need to be considered on the final activity. It's highly likely that putatively antihypertensive peptide sequences are subjected to alteration before the final activity *in vivo* after the various steps, such as attack of gastrointestinal enzymes and brush border peptidases, absorption through the intestinal barrier, attack of intracellular peptidases in the transcellular absorption and plasma enzymes after the peptides have entered the circulation (Vermeirssen et al., 2004; De Leo et al., 2009).

As human studies are time-consuming, costly and restricted by ethical concerns, the development of *in vitro* models for investigating the effects of digestion on the bioactive peptides has attracted much attention. A plethora of models have been implemented to simulate gastrointestinal digestion; most typically models simulate digestion in the oral cavity, the stomach and the small intestine. There is substantial variability among the conditions, such as the time of digestion, agitation, enzymes and concentrations of the salts and bile acids used. Moreover, there is great differentiation in the inclusion of various digestion stages and whether the chosen conditions are static with fixed concentrations of enzymes and bile acids, etc. or dynamic with varying concentrations of these constituents. For instance, human digestive liquids have been utilized to model digestion *in vitro* (Eriksen et al., 2010; Mäkinen et al., 2012), and meanwhile several reports have concerned the implementation of porcine enzyme mixtures (for example, Vermeirssen et al., 2003; Lo and Li-Chan, 2005; Marambe et al., 2011). The diversity of model conditions has hampered the ability to compare results across the different studies. Thus, a consensus concerning the basic parameters would be relevant in order to harmonize the various *in vitro* digestion models. Recently, *in vitro* models for studying the digestion of secondary plant metabolites (such as polyphenols) were reviewed by Alminger et al., (2014). A set of parameters for static *in vitro* models was suggested based on relevant *in vivo* data. Thus far, such a harmonization of parameters based on *in vivo* data has not been reported for modeling the bioavailability of peptides. Rapeseed protein Subtilisin digest and Alcalase digests have been subjected to simulated gastrointestinal digestion *in vitro* in order to elucidate the ability of ACE inhibitory peptides present in these digests to resist gastrointestinal digestion and the possibilities of reaching the site for absorption (Marczak et al., 2003; Mäkinen et al., 2012). The Alcalase hydrolysate was treated with human gastric extracts and Subtilisin hydrolysate with commercial proteases and results from both studies indicated that the ACE inhibitory peptides present in the hydrolysates possess a relatively high ability to resist further hydrolysis in to inactive degradation products (Mäkinen et al., 2012; Marczak et al., 2003).

Study of intestinal absorption *in vitro* is another common aim when elucidating the bioavailability. It has been indicated that a small portion of bioactive peptides can pass the intestine barrier and although it is usually too small to be considered nutritionally important, it can present the biological effects in tissue level (Gardner, 1988; Gardner, 1998). Molecular size and structural properties, such as hydrophobicity, affect the major transport route for peptides (Gardner, 1998). Research findings indicate that peptides with 2–6 amino acids are absorbed more readily in comparison to protein and free amino acids. As the molecular weight of peptides increases, their chance to pass the intestinal barrier decreases. Peptides are transported by active transcellular transport or by passive process (Shimizu et al., 1997). The absorption studies are commonly performed with the monolayer of intestinal cell lines, such as Caco-2 cells, simulating intestinal epithelium, and analysis of peptides and metabolites in serum after *in vivo* and clinical studies. Foltz et al., (2008) investigated the transport of Ile-Pro-Pro and Val-Pro-Pro by using three different absorption models and demonstrated that these tri-peptides are transported in small amounts intact across the barrier of the intestinal epithelium. In another study, the absolute bioavailability of the tri-peptides in pigs was below 0.1%, with an extremely short elimination half-life ranging from 5 to 20 min (van der Pijl et al., 2008). In humans, maximal plasma concentration did not exceed picomolar concentration (Foltz et al., 2007). Milk-derived peptide Leu-His-Leu-Pro-Leu-Pro is an interesting example of a peptide with evaluation of bioavailability. This peptide resisted gastrointestinal simulation, but cellular peptidases digested the peptide to His-Leu-Pro-Leu-Pro before crossing Caco-2 cell monolayer (Quiros et al., 2008; Quiros et al., 2009). The degradation product, His-Leu-Pro-Leu-Pro, has been demonstrated to absorb in human intestine as it has been identified in human plasma after oral administration (van Platerink et al., 2006).

Fujita and colleagues (Fujita et al., 2006) established a bioavailability factor in relation to antihypertensive activity and ACE-inhibition mechanism. The classification is based on inhibitor type and substrate type, the possible conversion of peptides by ACE into peptides with weaker activity and pro-drug type inhibitors, or possible conversion of peptides into true inhibitors by ACE or gastrointestinal proteases. A delayed antihypertensive effect is characteristic for pro-drug type peptides as they need to degrade further to reach the final active form (Zhao et al., 2009; Muguruma et al., 2009). For instance, flaxseed protein showed pro-drug type characteristics compared to hydrolysed cationic peptide fraction (Udenigwe et al., 2009). The protein fraction showed a delayed hypotensive effect in SHR comparable to captopril (3 mg/kg body weight) and the effect was more sustained than the effect of the digested peptide fraction. The slow-acting character of the protein fraction was expected since the digestion of the proteins. Anyhow, more research is needed to identify the active peptide sequences released in the digestive tract and to evaluate the bioavailability of these peptides.

It can be deduced due to the incomplete bioavailability of peptide following oral ingestion, a peptide with pronounced antioxidant activity *in vitro* may exert little or no activity *in vivo*. However, bypass routes which increase the chance of peptide absorption can diminish the problem and it is possible that *in vivo* antioxidant activity can be higher than *in vitro* activity. In such cases, bioactive peptides may display their biological functions by mechanisms other than what is applied in experiment. In addition, it has been suggested that the strong *in vivo* activity can be due to increased activity of peptides following their breakdown by gastrointestinal proteases (Erdmann et al., 2008).

The need for technological tools for the improvement of bioavailability is reinforced, and the focus is on the improvement of limited stability and absorption of peptides. The improvement of limited absorption and stability of peptides has been a goal when evaluating their effectiveness. For example, some carriers interact with the peptide molecule to create an insoluble entity at low pH, which later dissolves and facilitates intestinal uptake, by enhancing peptide transport over the non-polar biological membrane (Shaji and Patole, 2008). Bioavailability of bioactive tri-peptides (Val-Pro-Pro, Ile-Pro-Pro, Leu-Pro-Pro) was improved by administering them with a meal containing fibre, as compared to a meal containing no fibre. High methylated citrus pectin was used as a fibre (Kies and Van Der Pijl, 2012). Among drug delivery systems, emulsions have been used to enhance oral bioavailability or promoting absorption through mucosal surfaces of peptides and proteins (Shaji and

Patole, 2008). Individually, various components of emulsions have been considered as candidates for improving bioavailability of peptides. Anyhow, it seems that no general strategy for improving bioavailability of antihypertensive peptides exists and due to the number of processes involved and different characteristics of peptides depending on the sequence each case must be studied. Many strategies are currently demonstrated for enhancing bioavailability (Nestor, 2009) among them microencapsulation for controlled release of the active compounds, stabilization of the active molecules to improve transportation through the intestinal barrier and provide resistance against degradation, and development of highly stabile peptide analogues (Yamada et al., 2002; Matoba et al., 2001; Gomez-Guillen et al., 2011).

Incorporation into Food Products

Intensive research on bioactive peptides being carried out around the world has already led to the introduction of a wide range of commercial products. The bioactive peptides offer an exciting opportunity in the area of the development of novel functional foods which in turn could contribute to the prevention and management of certain diseases such as hypertension, Type 2 diabetes or obesity and more broadly metabolic syndrome. Antioxidant peptides could be most appropriate for specific populations, with increased risk of oxidative stress, such as ageing. Nutrient solutions that provide a rich source of essential amino acids and antioxidant peptides could provide a novel approach to promote healthy aging. The functional foods or food ingredients containing milk-derived bioactive peptides, such as the fermented milk Calpis, are already in the market (Phelan et al., 2009). The claims related to peptides are hypotensive properties, aiding mineral absorption, improving athletic performance and reduce stress. Since 1991, the Ministry of Health and Welfare in Japan has awarded the status of *Food of Specific Health Use* (FOSHU) to foods with scientifically validated health claims. Since then, antihypertensive peptides, such as Val-Pro-Pro, Ile-Pro-Pro, Val-Tyr have obtained FOSHU approval (Phelan et al., 2009). In Europe, applications for nutrition and health claims are submitted to the European Food Safety Authority (EFSA) under Regulation 1924/2006 and are evaluated by Dietetic Products, Nutrition and Allergies (NDA) panel of scientific experts (Anon 2006). There are three categories of health claims as defined by EU legislation. Article 13.1 claims are defined as new function or emerging science claims. Recently, the aspects concerning the scientific information needed for the use of a health claim in the functional food product labeling and marketing should include the scientific evidence on the beneficial effects of the product. The characterization of food components with *in vitro* and animal models is needed but they are not sufficient to substantiation the biological functionality in humans. Human studies to investigate the effects of food or food components on reliable markers, such as blood pressure and oxidative damage, are essential. There is still a lot of confusion within the food industry as to what evidence is required with the EU. Regarding the applications already processed, the Commission of European Communities has not yet authorized any claims relating to the effect of bioactive peptides in foods.

Future Prospects

There is no doubt that the hydrolysis of proteins gives rise to diversity of peptides, some of them displaying remarkable functionalities relevant to human health. The research should encourage the industry to invest more in the added-value products with scientific evidence of health benefits. To this end, novel technologies are available to standardize and stabilize the concentrations of active peptides in the products by means of membrane separation techniques and encapsulation. Important structure-function parameters of peptides are increasing constantly, which can greatly enhance the production and processing of peptides. With improved understanding of the structure activity relationship we may be able to design targeted enzyme hydrolysis strategies to release these peptides. Also, the integrity of the chemical structure of peptides, their interactions with food matrix components and organoleptic properties are important attributes to be considered when formulating novel products containing bioactive peptides.

Some studies have shown it to be a wrong approach that peptides are screened first *in vitro* for potential targets and then *in vivo* to confirm efficacy. According to Foltzand collagues (2010), it appears that it is only valid to propose efficacy once the peptide exhibits reasonable proteolytic stability and physiologically relevant absorption, distribution, metabolism and excretion profiles. In this field, more in-depth topics include the stability of the biological activity of peptides, both in the food matrix where released or incorporated and *in vivo* in the body before being absorbed and transported to the target site. Greater understanding of the biological fate of peptides and the site of action will allow delivery or an effective dose and formulation of the peptides to ensure that they reach their target sites. Confirmation of the efficacy of plant-derived peptides *in vivo* is still in its infancy. Some food protein-derived peptides and intact proteins have shown a beneficial effect *in vitro* and in human intervention studies. However, we need to gain better understanding of the relationship between these *in vitro* activities and especially, long term health benefits in humans and establish appropriate biomarkers of biological efficacy. For example, the extent of the antihypertensive effects has been suggested to depend on the nature of delivery system, dose, study duration, genetic background of the subjects, stages of hypertension (reviewed by Udenigwe and Aluko, 2012). Furthermore, molecular studies are needed to assess the mechanisms by which bioactive peptides exert their activities in the body. To this end, it may be necessary to employ proteomic and metabolomic methods. By means of these novel nutrigenomic approaches, it is possible and, in future, perhaps essential to investigate the impact of peptides on the expression of genes and hence endeavour to optimize the nutritional and health effects delivered by peptides.

Moreover, the safety of all novel peptides intended for food or pharmaceutical uses should be tested in accordance with international and national food safety regulations. In cases of products intended to be marketed in the EU member states, the novel food legislation has to be observed. Other challenges with dietary bioactive peptides are posed by health claims, which in the EU countries are strictly regulated and require science-based documentation before approval by the European Commission. At present, there are worldwide efforts to harmonize these regulations so as to develop fair global food marketing and protect consumers against false or misleading product information.

REFERENCES

Alashi, A. M., Blanchard, C. L., Mailer, R. J., Agboola, S. O., Mawson, A. J., He, R., et al., (2014). Blood pressure lowering effects of Australian canola protein hydrolysates in spontaneously hypertensive rats. *Food Res. Int.*, *55*, 281–287.

Alminger, M., Aura, A. M., Bohn, T., Dufour, C., El, S. N., Gomes, A., Karakaya, S., Martínez-Cuesta, M. C., McDougall, G. J., Requena, T. & Santos, C. N. (2014). *In vitro* Models for Studying Secondary Plant Metabolite Digestion and Bioaccessibility. *Compr. Rev. Food Sci. Food Safety*, *13*, 413–436.

Anderson, J. W. & Siesel, A. E. (1990). Hypocholesterolemic effects of oat products. In: New developments in dietary fiber: physiological, physicochemical, and analytical aspects. I. Furda and C. J. Brine (Eds.). New York, *Plenum Press.*, 17-36.

Anon, (2006). Regulation (EC) NO 1924/2006 of the European Parliament and of the Council of 20 December 20026 on Nutrition and Health Claims made on foods. *Off J Eur Union OJ L*, *12*, 3-18.

Barbana, C. & Boye, J. I. (2010). Angiotensin I-converting enzyme inhibitory activity of chickpea and pea protein hydrolysates. *Food Res. Int.*, *43*, 1642–1649.

Cao, G. & Prior, R. L. (1998). Comparison of different analytical methods for assessing total antioxidant capacity of human serum. *Clin. Chem.*, *44*, 1309–1315.

Carmel, A. & Yaron, A. (1978). An intramolecularly quenched fuorescent tripeptide as a fuorogenic substrate of angiotensin-I-converting enzyme and of bacterial dipeptidyl carboxypeptidase. *Eur. J. Biochem.*, *87*, 265.

Chiang, W. E., Cordle, C. T. & Thomas, R. L. (1995). Casein hydrolysate produced using a formed-in-place membrane reactor. *J. Food Sci.*, *60*, 1349-1352.

Cushman, D. W. & Cheung, H. S. (1971). Spectrophotometric assay and properties of the angiotensin converting enzyme of rabbit lung. *Biochem. Pharmacol.*, *20*, 1637-1648.

De Leo, F., Panarese, S., Gallerani, R. & Ceci, L. R. (2009). Angiotensin converting enzyme (ACE) inhibitory peptides: production and implementation of functional food. *Curr. Pharm. Design*, *15*, 3622-3643.

Doig, M. T. & Smiley, J. W. (1993). Direct injection assay of angiotensin-converting enzyme by high-performance liquid-chromatography using a shielded hydrophobic phase column. *J. Chrom. Biomed. Appl.*, *613*, 145-149.

Doyen, A., Beaulieu, L., Saucier, L., Pouliot, Y. & Bazinet, L. (2011b). Demonstration of *in vitro* anticancer properties of peptide fractions from a snow crab by-products hydrolysate after separation by electrodialysis with ultrafiltration membranes. *Sep. Purif. Technol.*, *78*, 321–9.

Doyen, A., Beaulieu, L., Saucier, L., Pouliot, Y. & Bazinet, L. (2011a). Impact of ultrafiltration membrane material on peptide separation from a snow crab byproduct hydrolysate by electrodialysis with ultrafiltration membranes. *J. Agric. Food Chem.*, *59*, 1784–92.

Elisia, I. & Kitts, D. D. (2008). Anthocyanins inhibit peroxy radical-induced apoptosis in Caco-2 cells. *Mol. Cell. Biochem.*, *312*, 139–145.

Erdmann, K., Cheung, B. W. & Schröder, H. (2008). The possible roles of food-derived bioactive peptides in reducing the risk of cardiovascular disease. *J. Nutr. Biochem.*, *19*, 643-654.

Eriksen, E. K., Holm, H., Jensen, E., Aaboe, R., Devold, T. G., Jacobsen, M., et al., (2010). Different digestion of caprine whey proteins by human and porcine gastrointestinal enzymes. *Br. J. Nutr.*, *104*, 374-381.

Firdaous, L., Dhulster, P., Amiot, J., Gaudreau, A., Lecouturier, D., Kapel, R., et al., 2009. Concentration and selective separation of bioactive peptides from an alfalfa white protein hydrolysate by electrodialysis with ultrafiltration membranes. *J. Memb. Sci.*, *329*, 60–67.

Fitzgerald, C., Gallagher, E., Tasdemir, D. & Hayes, M. (2011). Heart health peptides from macroalgae and their potential use in functional foods. *J. Agric. Food Chem.*, *59*, 6829-6836.

Fitzgerald, R. J., Murray, B. A. & Walsh, D. J. (2004). Hypotensive peptides from milk proteins. *J. Nutr.*, *134*, 980-988.

Foltz, M., Cerstiaens, A., van Meensel, A., Mols, R., van der Pijl, P. C., Duchateau, G. S. & Augustijns, P. (2008). The angiotensin converting enzyme inhibitory tripeptides Ile-Pro-Pro and Val-Pro-Pro show increasing permeabilities with increasing physiological relevance of absorption models. *Peptides*, *29*, 1312–1320.

Foltz, M., Meynen, E. E., Bianco, V., van Platerink, C., Koning, T. M. & Kloek, J. (2007). Angiotensin converting enzyme inhibitory peptides from a lactotripeptide-enriched milk beverage are absorbed intact into the circulation. *J. Nutr.*, *137*, 953–958.

Fujita, H., Yokoyama, K. & Yoshikawa, M. (2000). Classification and antihypertensive activity of angiotensin I-converting enzyme inhibitory peptides derived from food proteins. *J. Food Sci.*, *65*, 564-569.

Gardner, M. L. G. (1988). Gastrointestinal absorption of intact proteins. *Annu. Rev. Nutr.*, *8*, 329–350.

Gardner, M. L. G. (1998). Transmucosal passage on intact peptides in mammalian metabolism. In: Grimble GK, Backwell FRG, editors. Tissue utilization and clinical targeting. London: Portland Press Ltd.

Gomez-Guillen, M. C., Gimenez, B., Lopez-Cabacllero, M. E. & Montero, M. P. (2011). Functional and bioactive properties of collagen and gelatin from alternative sources: A review. *Food Hydrocolloids*, *25*, 1813-1827.

Guang, C. & Phillips, R. D. (2009). Plant food-derived angiotensin I converting enzyme peptides. *J. Agric. Food Chem.*, *57*, 5113-5120.

Guérard, F. (2007). Enzymatic methods for marine by-products recovery. In: Shahidi F editor. Maximizing the value of marine by-products. Campridge: Woodward Publishing Limited pp. 107–143.

Hartmann, R. & Meisel, H. (2007). Food-derived peptides with biological activity from research to food applications. *Curr. Opin. Biotechnol.*, *18*, 163-169.

He, R., Alashi, A., Malomo, S. A., Girgih, A. T., Chao, D., Ju, X., et al., (2013a). Antihypertensive and free radical scavenging properties of enzymatic rapeseed protein hydrolysates. *Food Chem.*, *141*, 153–159.

He, R., Malomo, S. A., Alashi, A., Girgih, A. T., Jua, X., Aluko, R. E. (2013b). Purification and hypotensive activity of rapeseed protein-derived renin and angiotensin converting enzyme inhibitory peptides. *J. Funct. Foods*, *5*, 781-9.

He, R., Malomo, S. A., Girgih, A. T., Ju, X. & Aluko, R. E. (2013c). Glycinyl-Histidinyl-Serine (GHS), a novel rapeseed protein-derived peptide has blood pressure-lowering effect in spontaneously hypertensive rats. *J. Agric. Food Chem.*, *61*, 8396–402.

Hernández-Ledesma, B., Miralles, B., Amigo, L., Ramos, M. & Recio, I. (2005). Identification of antioxidant and ACE-inhibitory peptides in fermented milk. *J. Sci. Food Agric.*, *85*, 1041-1048.

Holmquist, B., Bunning, P. & Riordan, J. (1979). Continuous spectrophotometric assay for angiotensin converting enzyme. *Anal. Biochem.*, *95*, 540-548.

Inouye, K., Nakano, K., Asaoka, K. & Yasukawa, K. (2009). Effects of thermal treatment on the coagulation of soy proteins induced by Subtilisin Carlsberg. *J. Agric. Food Chem.*, *57*, 717-23.

Kies, A. K. & Van Der Pijl, P. (2012). Peptide availability. USA Patent Application 20120040895.

Lavoie, J. L. & Sigmund, C. D. (2003). Minireview: Overview of the renin-angiotensin system -An endocrine and paracrine system. *Endocrinol.*, *144*, 2179-2189.

Li, G. H., Liu, H., Shi, Y. H. & Le, G. W. (2005). Direct spectrophotometric measurement of angiotensin I-converting enzyme inhibitory activity for screening bioactive peptides. *J. Pharm. Biomed. Anal.*, *37*, 219224.

Liu, R. H. & Finley, J. (2005). Potential cell culture models for antioxidant research. *J. Agric. Food Chem.*, *53*, 4311–4314.

Lo, W. M. Y. & Li-Chan, E. C. Y. (2005). Angiotensin I converting enzyme inhibitory peptides from *in vitro* Pepsin-Pancreatin digestion of soy protein. *J. Agric. Food Chem.*, *53*, 3369-3376.

Ma, Y. & Xiong, Y. L. (2009). Antioxidant and bile acid binding activity of buckwheat protein *in vitro* digests. *J. Agric. Food Chem.*, *57*, 4372-4380.

Mäkinen, S., Johansson, T., Vegarud, G., Pihlava, J. M. & Pihlanto, A. (2012). Angiotensin Iconverting enzyme inhibitory and antioxidant properties of rapeseed hydrolysates. *J. Funct. Foods*, *4*, 575-583.

Mannheim, A. & Cheryan, M. (1990). Continuous hydrolysis of milk protein in a membrane reactor. *J. Food Sci.*, *55*, 381-385.

Marambe, H. K., Shand, P. J. & Wanasundara, J. P. D. (2011). Release of angiotensin I-converting enzyme inhibitory peptides from flaxseed (*Linum usitatissimum* L.) protein under simulated gastrointestinal digestion. *J. Agric. Food Chem.*, *59*, 9596-9604.

Marczak, E. D., Ohinata, K., Lipkowski, A. W. & Yoshikawa, M. (2006). Arg-Ile-Tyr (RIY) derived from rapeseed protein decreases food intake and gastric emptying after oral administration in mice. *Peptides*, *27*, 2065-2068.

Marczak, E. D., Usui, H., Fujita, H., Yang, Y., Yokoo, M., Lipkowski, A. W., et al., (2003). New antihypertensive peptides isolated from rapeseed. *Peptides*, *24*, 791–798.

Mariotti, F., Hermier, D., Sarrat, C., Magné, J., Fenart, E., Evrard, J., et al., (2008). Rapeseed protein inhibits the initiation of insulin resistance by a high-saturated fat, high-sucrose diet in rats. *Br. J. Nutr.*, *100*, 984-991.

Matoba, N., Yamada, Y., Usui, H., Nakagiri, R. & Yoshikawa, H. (2001). Designing potent derivatives of ovokinin (2-7), an anti-hypertensive peptide derived from ovalbumin. *Biosci. Biotechn. Biochem.*, *65*, 736-739.

Megias, C., Pedroche, J., Yust, M. D. M., Alaiz, M., Giron-Calle, J., Millan, F., et al., (2006). Affinity purification of angiotensin converting enzyme inhibitory peptides using immobilized ACE. *J. Agric. Food Chem.*, *54*, 7120-7124.

Muguruma, M., Ahhmed, A. M., Katayama, K., Kawahara, S., Maruyama, M. & Nakamura, T. (2009). Identification of pro-drug type ACE inhibitory peptide sourced from porcine myosin B: Evaluation of its antihypertensive effects *in vivo*. *Food Chem.*, *114*, 516-522.

Nestor, J. J. Jr. (2009). The medicinal chemistry of peptides. *Curr. Med. Chem.*, *16*, 4399-4418.

Orsini-Delgado, M. C., Tironi, V. A. & Anon, C. M. (2011), Antioxidant activity of amaranth protein or their hydrolysates under simulated gastrointestinal digestion. *LWT-Food Sci. Techn.*, *44*, 1752–1760.

Pedroche, J., Yust, M. M., Giron-Calle, J., Alaiz, M., Millan, F. & Vioque, J. (2002). Utilisation of chickpea protein isolates for production of peptides with angiotensin I-converting enzyme (ACE)-inhibitory activity. *J. Sci. Food. Agric.*, *82*, 960-965.

Phelan, M., Aherne, A., FitzGerald, R. J. & O'Brien, N. M. (2009). Casein-derived bioactive peptides: Biological effects, industrial uses, safety aspects and regulatory status. *Int. Dairy J.*, *19*, 643-654.

Pihlanto, A., Johansson, T. & Mäkinen, S. (2012). Inhibition of angiotensin I-converting enzyme and lipid peroxidation by fermented rapeseed and flaxseed meal. *Eng. Life Sci.*, *12*, 450-456.

Pihlanto, A. & Korhonen, H. (2003). Bioactive peptides and proteins. *Adv. Food Res.*, *47*, 175-276.

Pihlanto, A., Virtanen, T. & Korhonen, H. (2010). Angiotensin I converting enzyme (ACE) inhibitory activity and antihypertensive effect of fermented milk. *Int. Dairy J.*, *20*, 3-10.

Pownall, T. L., Udenigwe, C. C. & Aluko, R. E. (2010). Amino acid composition and antioxidant properties of pea seed (Pisum sativum L.) enzymatic protein hydrolysate fractions. *J. Agric. Food Chem.*, *58*, 4712–4718.

Pownall, T. L., Udenigwe, C. C. & Aluko, R. E. (2011). Effects of cationic property on the *in vitro* antioxidant activities of pea protein hydrolysate fractions. *Food Res. Int.*, *44*, 1069–74.

Quiros, A., Davalos, A., Lasuncion, M. A., Ramos, M. & Recio, I. (2008). Bioavailability of the antihypertensive peptide LHLPLP: Transepithelial flux of HLPLP. *Int. Dairy J.*, *18*, 279-286.

Quiros, A., del Mar Contreras, M., Ramos, M., Amigo, L. & Recio, I. (2009). Stability to gastrointestinal enzymes and structure-activity relationship of beta-casein-peptides with antihypertensive properties. *Peptides*, *30*, 1848-1853.

Re, R., Pellegrini, N., Proteggente, A., Pannala, A., Yang, M. & Rice- Evans, C. (1999). Antioxidant activity applying an improved ABTS radical cation decolorization assay. *Free Radical Bio. Med.*, *26*, 1231–1237.

Scheraga, H. A. (2004). The thrombin–fibrinogen interaction. *Biophys. Chem.*, *112*, 117–130.

Sentandreu, M. A. & Toldrá, F. (2006). A rapid, simple and sensitive fluorescence method for the assay of angiotensin-I converting enzyme. *Food Chem.*, *97*, 546-554.

Shaji, J. & Patole, V. (2008). Protein and peptide drug delivery: Oral approaches. *J. Pharm. Sci.*, *70*, 269-277.

Shalaby, S. M., Zakora, M. & Otte, J. (2006). Performance of two commonly used angiotensin converting enzyme inhibition assays using synthetic peptide substrates. *J. Dairy Res.*, *73*, 178-186.

Shimizu, M., Tsunogai, M. & Arai, S. (1997). Transepithelial transport of oligopeptides in the human intestinal cell, Caco2. *Peptides*, *18*, 681–687.

Silva, S. V. & Malcata, F. X. (2005). Caseins as source of bioactive peptides. *Int. Dairy J.*, *15*, 1–15.

Udenigwe, C., Lin, Y. S., Hou, W. C. & Aluko, R. E. (2009). Kinetics of the inhibition of rennin and angiotensin I-converting enzyme by flaxseed protein hydrolysate fractions. *J. Funct. Foods*, *1*, 199-207.

Udenigwe, C. & Aluko, R. (2012). Food protein-derived bioactive peptides: Production, processing, and potential health benefits. *Concise Rev. Food Sci.*, *71*, R11-R24.

Udenigwe, C. C. & Aluko, R. E. (2010). Antioxidant and angiotensin converting enzyme-inhibitory properties of a flaxseed protein derived high Fischer ratio peptide mixture. *J. Agric. Food Chem.*, *58*, 4762–4768.

Udenigwe, C. C., Adebiyi, A. P., Doyen, A., Li, H., Bazinet, L. & Aluko, R. E. (2012). Low molecular weight flaxseed protein-derived arginine-containing peptides reduced blood pressure of spontaneously hypertensive rats faster than amino acid form of arginine and native flaxseed protein. *Food Chem.*, *132*, 468-475.

van der Pijl, P. C., Kies, A. K., Ten Have, G. A., Duchateau, G. S. & Deutz, N. E. (2008). Pharmacokinetics of proline-rich tripeptides in the pig. *Peptides*, *29*, 2196–2202.

van Platerink, C., Janssen, H., Horsten, R. & Haverkamp, J. (2006). Quantification of ACE inhibiting peptides in human plasma using high performance liquid chromatography mass spectrometry. *J. Chromatogr. B*, *830*, 151-157.

Verma, S., Buchanan, M. R. & Anderson, T. J. (2003). Endothelial function testing as a biomarker of vascular disease. *Circulation*, *108*, 2054-2059.

Vermeirssen, V., Van Camp, J., Verstraete, A. & Verstraete, W. (2004). Bioavailability of angiotensin I converting enzyme inhibitory peptides. *Br. J. Nutr.*, *92*, 357-366.

Vermeirssen, V., Augustijns, P., Morel, N., Van Camp, J., Opsomer, A. & Verstraete, W. (2005). *In vitro* intestinal transport and antihypertensive activity of ACE inhibitory pea and whey digests. *Int. J. Food Sci. Nutr.*, *56*, 415-430.

Vermeirssen, V., van Camp, J., Decroos, K., van Wijmelbeke, L. & Verstaete, W. (2003). The impact of fermentation and *in vitro* digestion on the formation of angiotensin-I-converting enzyme inhibitory activity from pea and whey protein. *J. Dairy Sci.*, *56*, 429–438.

Vermeirssen, V., Van Camp, J. & Verstraete, W. (2002). Optimisation and validation of an angiotensin-converting enzyme inhibition assay for the screening of bioactive peptides. *J. Biochem. Biophys. Methods*, *51*, 75-87.

Virtanen, T., Pihlanto, A., Akkanen, S. & Korhonen, H. (2007). Development of antioxidant activity in milk whey during fermentation with lactic acid bacteria. *J. Appl. Microbiol.*, *102*, 106–115.

Vuorela, S., Meyer, A. S. & Heinonen, M. (2004). Impact of isolation method on the antioxidant activity of rapeseed meal phenolics. *J. Agric. Food Chem.*, *52*, 8202–8207.

Wang, Y. K., He, H. L., Wang, G. F., Wu, H., Zhou, B. C., Chen, X. L. & Zhang, Y. Z. (2010). Oyster (Crassostrea gigas) hydrolysates produced on a plant scale have antitumor activity and immunostimulating effects in BALB/c mice. *Mar. drugs.*, *8*, 255–268.

WHO World Health Organization. (2011). Cardiovascular Diseases (CVD's). Fact sheet N°317.

Wolfe, K. L. & Liu, R. H. (2007). Cellular antioxidant activity (CAA) assay for assessing antioxidants, foods, and dietary supplements. *J. Agric. Food Chem.*, *55*, 8896–8907.

Wu, J., Aluko, R. E. & Muir, A. D. (2008). Purification of angiotensin I-converting enzyme-inhibitory peptides from the enzymatic hydrolysate of defatted canola meal. *Food Chem.*, *111*, 942-950.

Wu, J. & Ding, X. (2002). Characterization of inhibition and stability of soy-protein-derived angiotensin I-converting enzyme inhibitory peptides. *Food Res. Int.*, *35*, 367-375.

Xue, Z., Yu, W., Liu, Z., Wu, M., Kou, X. & Wang, J. (2009a). Preparation and antioxidative properties of a rapeseed (*Brassica napus*) protein hydrolysate and three peptide fractions. *J. Agric. Food Chem.*, *57*, 5287–5293.

Xue, Z., Yu, W., Wu, M. & Wang, J. (2009b). *In vivo* antitumor and antioxidative effects of a rapeseed meal protein hydrolysate on an S180 tumor-bearing murine model. *Biosci. Biotechnol. Biochem.*, *73*, 2412-2415.

Yamada, Y., Matoba, N., Usui, H., Onishi, K. & Yoshikawa, M. (2002). Design of a highly potent anti-hypertensive peptide based on ovokinin (2-7). *Biosci. biotechn. biochem.*, *66*, 1213-1217.

Yamada, Y., Iwasaki, M., Usui, H., Ohinata, K., Marczak, E. D., Lipkowski, A. W., et al., (2010). Rapakinin, an anti-hypertensive peptide derived from rapeseed protein, dilates mesenteric artery of spontaneously hypertensive rats via the prostaglandin IP receptor followed by CCK1 receptor. *Peptides*, *31*, 909-914.

Yoshie-Stark, Y., Wada, Y. & Waesche, A. (2008). Chemical composition, functional properties, and bioactivities of rapeseed protein isolates. *Food Chem.*, *107*, 32-39.

Yoshie-Stark, Y. & Waesche, A. (2004). *In vitro* binding of bile acids by lupin protein isolates and their hydrolysates. *Food Chem.*, *88*, 179-184.

Yu, W., Gao, J., Xue, Z., Kou, X., Wang, Y. & Zhai, L. (2013). Radical-scavenging activity, ACE-inhibiting capability and identification of rapeseed albumin hydrolysate. *Food Science and Human Wellness*, *2*, 93–98.

Yust, M. M., Pedroche, J., Giron-Calle, J., Alaiz, M., Francisco, M. & Vioque, J. Production of ace inhibitory peptides by digestion of chickpea legumin with Alcalase. *Food Chem.*, *81*, 363–369.

Zhang, L., Li, J. & Zhou, K. (2010). Chelating and radical scavenging activities of soy protein hydrolysates prepared from microbial proteases and their effect on meat lipid peroxidation. *Bioresource Technol.*, 2010, *101*, 2084–2089.

Zhang, S. B., Wang, Z. & Xu, S. Y. (2008). Antioxidant and antithrombotic activities of rapeseed peptides. *J. Am. Oil Chem. Soc.*, *85*, 521-527.

Zhang, S. B., Wang, Z., Xu, S. Y. & Gao, X. F. (2009). Purification and characterization of a radical scavenging peptide from rapeseed protein hydrolysates. *J. Am. Oil Chem. Soc.*, *86*, 959-966.

Zhao, Y, Li, B., Dong, S., Liu, Z., Zhao, X., Wang, J. & Zeng, M. (2009). A novel ACE inhibitory peptide isolated from Acaudina molpadioidea hydrolysate. *Peptides*, *30*, 1028-1033.

In: Rapeseed
Editor: Monica White

ISBN: 978-1-63484-227-3

Chapter 2

ENZYME-AIDED PROCESSING OF RAPESEED OIL INDUSTRY BY-PRODUCTS INTO ADDED VALUE INGREDIENTS FOR FOOD, COSMETIC AND PHARMACEUTICAL APPLICATIONS

***Diana Rivera*[1], *Katariina Rommi*[2], *Margarida M. Fernandes*[1], *Raija Lantto*[2] and *Tzanko Tzanov*[1,*]**

[1]Grup de Biotecnologia Molecular i Industrial, Department d'Enginyeria Química, Universitat Politècnica de Catalunya, Barcelona, Spain
[2]VTT Technical Research Centre of Finland Ltd, Finland

ABSTRACT

Rapeseed is the traditional name of a large winter or spring annual oilseed crop from the *Brassicaceae* family, and is related to mustard, cabbage, broccoli, among others. As one of the leading oilseed crops, rapeseed is primarily grown for its oil content. Production of rapeseed oil by cold or hot screw pressing solely or followed by solvent extraction, respectively, generates a significant co-stream in the form of press cake

[*] Corresponding author: Dr. Tzanko Tzanov. Grup de Biotecnologia Molecular i Industrial, Department d'Enginyeria Química, Universitat Politècnica de Catalunya, Rambla Sant Nebridi 22, 08222, España. Tel.: +34 93 739 85 70; fax: +34 93 739 82 25; e-mail: tzanko.tzanov@upc.edu.

or meal. This co-stream is an excellent feedstock for nutritionally valuable components such as protein, and bioactive compounds such as phenolics and peptides, not only for livestock but also for human consumption. The mentioned bioactive compounds have been found to possess potential health benefits and skin-repairing properties and thus may play an important role not only in the cosmetics but also in the treatment or prevention of many diseases. Rapeseed protein, constituting 32-36% of the rapeseed press cake or meal, has a well-balanced amino acid composition and suitable technological properties for food use. However, there are still challenges that limit the use of rapeseed press cake for commercial profitable production of protein and bioactive compounds, particularly in relation to the organoleptic properties and levels of harmful or anti-nutritional factors such as glucosinolates and phytates. This chapter reviews the possibilities as well as the challenges that rapeseed press cake co-stream ingredients face regarding skin care, food and nutraceutical applications.

INTRODUCTION

Rapeseed, which comprises the species *Brassica napus* and *Brassica rapa*, is among the most important oilseed crops which are grown annually in temperate climates such as Europe, Canada, China, India and Australia [1, 2]. The seeds (containing 40-48% oil on dry matter basis) are currently processed to produce oil that is sold mainly to the food and biodiesel market, but it is also suitable as a raw material for lubricants, plastics, soaps, etc. Depending on the method of oil extraction, the residual seed mass can be designated as: i) the press cake, which is produced when only physical oil pressing is used, or ii) the meal, which is obtained when the pressing is followed by solvent extraction of residual oil from the press cake [3].

The continuous growth of rapeseed oil production for food and other uses coincides with proportional production of co-streams. Global rapeseed press cake/meal production increased by 80% between 2002 and 2012 [4]. Similarly, in the EU-27, rapeseed press cake/meal production doubled from 6,5 million tons in 2004 to 13,5 million tons in 2014 [5]. Large processing plants use hot pressing and solvent extraction to obtain rapeseed oil, while smaller producers of unrefined virgin oil mainly adopt the cold pressing process. There is a trend among consumers towards the use of unrefined virgin oil particularly in Europe and Northern America. This multiplies the small- and medium-sized extraction plants producing virgin oil by cold pressing and consequently generating significant press cake co-stream.

After processing rapeseed into oil, many valuable compounds such as proteins, carbohydrates, lipids and a range of other compounds such as polyphenols, phenolic acids and esters, glucosinolates, phytates, vitamins and minerals remain in the press cake/meal. The chemical composition and nutritional value of these co-streams are influenced by the processing technology and conditions used in the oil production such as type of the oil press, temperature, humidity during the extraction process, etc.) [6], but also by environmental factors like climate conditions, soil and fertilisation, not to mention genetic variation caused by breeding [7]. These conditions set the potential and limitations of the obtained co-streams for use in high-value applications [8]. In particular, the press cake from cold pressing of rapeseed is considered as a raw material for food protein production since protein quality is better retained in this feedstock than in the solvent-extracted meal.

Current strategies for managing the co-streams include the use of press cake/meal as a protein source in feed formulations for animals, but also in lower value end-uses such as fertilizers or as a soil cover to prevent erosion. However, especially the press cake from cold pressing represents a rich source of native, high-quality protein and other bioactive compounds valuable as ingredients in foods, pharmaceuticals and cosmetics [9–14]. The press cake from cold pressing of whole rapeseeds contains 32-36% protein [15, 16], which is mainly comprised of the storage proteins cruciferin (12S globulin) and napin (2S albumin). The overall amino acid composition of these proteins is well-balanced with sufficient amounts of lysine and sulfur amino acids to meet the FAO requirements for daily intake [17, 18]. On the other hand, whole rapeseed protein seems to have limited digestibility in humans similarly to many other plant proteins [18, 19]. The functional properties (solubility, thermal, emulsification, foaming and gelation properties) of rapeseed proteins have been well characterized with clear indications of their potential as performance ingredients in e.g., food and cosmetic products [20]. Additionally, several bioactivities including antioxidative, antihypertensive, antidiabetic, anorexigenic, hypocholesterolemic, anticancer and antiviral properties, have been identified in rapeseed proteins and peptides [21].

Despite the larger potential of rapeseed press cake, there are still technological barriers to its full profitable use. The presence of phytates, which are efficient chelators of minerals and proteins [22], phenolic compounds generating undesirable colours and flavours, traces of glucosinolates which are considered antinutritional for animals, but have a controversial role in human health [23], and the high carbohydrate content (36% on dry matter basis) which influences the digestibility and technological properties of the press cake [16]. Therefore, feasible technologies are needed in order to enhance the functionality, bioavailability, and sensory properties of rapeseed-derived ingredients both in food and skin care applications. Aqueous extraction of proteins in alkaline or saline conditions has been actively studied since the 1970's with the aim to produce high-protein fractions for food use [24–26]. In addition, enzyme-aided processing using proteases and/or cell wall degrading enzymes has proven to be successful for the extraction of whole protein and protein hydrolysates from press cake and meal [16, 27–29].

The present chapter summarizes recent advances on the enzyme-assisted extraction processes of protein and other bioactive compounds from rapeseed press cake. It also reviews the potential use of these compounds in high value applications such as skin care, food and nutraceutical industries.

Sustainable Processing of Rapeseed Co-Streams

Oilseed proteins are most commonly enriched by alkaline extraction of defatted seed mass, followed by acidic precipitation and washing to yield a protein concentrate (>65% protein) or isolate (>90% protein). Although nearly pure protein can be obtained at a high yield, these technologies suffer from intensive water and energy consumption, use of chemicals, and requirement for multiple processing steps. High pH has been reported to expose proteins to denaturation, which reduces their solubility and related functional properties [30], and to increase the co-extraction of phenolic compounds with the protein [31]. To improve the feasibility, sustainability and economics of rapeseed protein production, alternative technologies such as dry fractionation and enzyme-assisted wet fractionation at reduced water content, have been the focus of several recent studies [16, 27–29, 32–36].

Dry fractionation is based on mechanical grinding and separation techniques and offers a sustainable approach for production of plant protein concentrates [33]. Dry fractionation is generally more energy-efficient than wet processing as there is no need for drying steps after the fractionation. Dry fractionation can also be regarded as a relatively mild way of processing since, although compromising the protein purity, it retains the natural functionality of the protein better than wet processing which is often carried out in severe pH and temperature conditions. Dry fractionation has been successfully used to enrich protein from pulses and cereals [33].

In rapeseed, protein and oil are mainly stored in the kernel of the seed which is surrounded by a hull rich in fibres and polyphenolics [37]. These components can have unfavourable effects on the sensory properties and digestibility of rapeseed protein products, and hence, removal of the hulls is recommended when targeting colour and flavour-neutral protein for food applications [38]. In our recent study, defatting of rapeseed cold-press cake by supercritical CO_2 extraction and subsequent removal of the hull particles by milling and air classification produced a fine, light-coloured material with concomitant increase of protein content from 36 to 46% [32]. Removal of oil from the cold-press-cake is a required pre-treatment and at the same time the major drawback of dry fractionation, since oil hinders the dispersion of milled particles during air classification [33].

Carbohydrate-hydrolysing enzymes, especially pectinase, cellulase, β-glucanase and xylanase, have proven useful for enrichment of rapeseed proteins. Carbohydrates constitute around 35% of rapeseed press cake [16], with pectin, cellulose, xyloglucan, arabinan and arabinogalactan being the main cell wall polysaccharides [39]. In the study by Niu et al., 2012, combined use of hemicellulase, pectinase, cellulase and protease enzymes enabled aqueous extraction of 82% of the total protein from rapeseed cold-press cake [28]. Such yield compares well with reported protein extraction yields from alkaline procedures, ranging from 56 [40] to 94% [41]. A patent by DeCastro et al., 2005, similarly describes a procedure where intermittent wet milling and enzymatic carbohydrate hydrolysis are used to recover a protein extract and oil-protein emulsion from thick slurry (19% solid content) of rapeseed cake and water [29].

Cell walls have been identified as one of the main physical barriers for protein release from various biomasses [42]. Pressing oil from rapeseeds partially disrupts seed cell walls and releases some protein from the kernel cells [16]. Cell wall degrading enzymes are expected to facilitate protein extraction from the press cake partly by disintegration of the remaining tissue [16]. Recently, we have revealed more mechanisms by which these enzymes influence rapeseed protein recovery, including the reduction of water-holding capacity of the press cake through pectin hydrolysis, and release of anionic compounds which improve protein solubility by electrostatic stabilization [34]. Enzyme treatment seems particularly beneficial when protein extraction is performed at increased solid content [34]. In these conditions, efficiency of solid-liquid separation by e.g., centrifugation is hindered by the high water holding capacity of rapeseed press cake, and carbohydrate-hydrolysing enzymes offer an effective tool to facilitate the processability of this material.

In addition to facilitating protein extraction from the press cake without alkali or other chemicals, carbohydrate-hydrolysing enzyme treatment has also shown to improve functional properties of the resulting protein extracts [35]. Generally, use of these enzymes results in increased co-extraction of sugars and galacturonic acid [16]. Anionic carbohydrates are known to stabilize rapeseed protein dispersions by formation of electrostatic complexes [43, 44], and in a similar manner, uronic acids released during pectin hydrolysis have been suggested to increase the dispersion stability of enzymatically produced crude protein extracts [35].

Protein-hydrolysing enzymes i.e., proteases have been increasingly studied in the past decade as a gentle method for production of rapeseed protein hydrolysates with new functional properties. Hydrolysis of rapeseed proteins into soluble peptides can substantially facilitate their recovery from the press cake and solvent-extracted meal [27, 28]. The degree of hydrolysis influences not only the yield but also the functional and sensory properties of the resulting peptides. Limited hydrolysis can improve the emulsifying and foaming activities and water and oil absorption capacities of rapeseed protein hydrolysates [12], while extensive hydrolysis may result in weakened functional properties [45] and formation of peptides with bitter taste [46]. On the other hand, small peptide size has been connected with certain bioactivities such as the inhibition of the angiotensin converting enzyme (ACE) involved in blood pressure regulation [47].

Various commercial protease preparations have proven effective for production of mixtures of bioactive rapeseed peptides. To maximize the yield of peptides, endopeptidases which cleave the peptide chains in the middle have been favoured over exopeptidases which release amino acids from peptide chain ends. The most studied preparations include *B. licheniformis* serine endopeptidases and mammalian gastrointestinal proteases [48].

Valorisation of Rapeseed Co-Streams into Added-Value Applications

Increasing global demand of proteins for human consumption and animal feeding has prompted the exploration of rapeseed industry co-streams, among others, as a novel protein source. Traditionally, rapeseed press cake/meal is used as animal feed, but its protein- and phenolic-rich nature, and the already proven bioactive properties of these compounds have found applications in other markets. Apart from food applications, their use as a resource for nutraceuticals and cosmeceuticals has been considered like various non-food applications such as films and paper additives [49]. Furthermore, it has been considered for its potential use in fermentative processes, including saccharification and production of enzymes, antibiotics, antioxidants and other by-products [50].

Annually, nearly 15 million tonnes of rapeseed co-streams are produced in Europe. The market price of press cake and meal used for animal feed is EUR 150-300/ton and the price of rapeseed oil varies between EUR 300 and 1400/ton [4]. In comparison, the price of plant protein ingredients such as soy protein isolates and hydrolysates ranges from 6000 to 7000 USD/ton [51], and the commercial value of food-quality rapeseed protein products made of press cakes could reach a similar level. So far, commercial value has not been estimated for the use of rapeseed press cake in cosmetics but an even higher value would be foreseen. Despite of these attractive price estimations, a small number of applications of the rapeseed co-streams are found in literature. Nevertheless, their potential use in added-value applications in food and cosmetic industry is a growing research interest.

New and valuable compounds can be extracted and purified from rapeseed co-streams using technologies of varying complexity. One of the early rapeseed protein concentrates, generated after a dehulling process followed by a myrosinase inactivation, was reported to possess an acceptable light colour, neutral and bland flavour, and high nutritional value [52]. According to later studies, rapeseed protein isolates have shown favourable functional properties such as high emulsifying capacity [14] or gelation efficiency [53] depending on the production method and composition. It is important to consider that although certain non-protein compounds such as phenolics can impair protein performance [54], high purity is not a prerequisite for suitable functional properties. In a recent study, a protein-rich fraction (46% protein) obtained by dry fractionation of defatted rapeseed press cake possessed lighter colour and higher dispersion stability than a rapeseed protein isolate (>90% protein) from wet fractionation [55].

The potential of rapeseed proteins in several food applications has been studied, though the brown colour of protein fractions, resulting from oxidation of phenolic compounds, is considered as a major limitation for their incorporation in food and other products [36, 48]. For example, rapeseed protein concentrates have been successfully applied as alternative for soy or lupine protein concentrates in sausage formulations to prevent water and oil separation [36, 48], and in bakery products and mayonnaise for complete substitution of egg protein.

Yoshie-Stark et al., 2006 used a steaming process to obtain rapeseed protein concentrates for food applications [48]. When this protein concentrate was added to sausage preparations to replace casein, sensory analysis showed that the rapeseed protein sausage had improved taste, good texture and a characteristic aroma. Moreover, after hydrolysis with gastrointestinal enzymes, an improved angiotensin converting enzyme (ACE) inhibition and DPPH radical scavenging activity was observed [56].

Apart from the protein, rapeseed press cake has also attracted much attention as a source of antioxidative phenolic compounds such as sinapine and sinapic acid which are enriched in the kernel of the seeds [57]. Rapeseed phenolic extracts have proved effective in preventing unwanted oxidation of emulsions and oils [58, 59], therefore they can serve as natural preservatives in various food and cosmetic products. More recently, a rapeseed press cake valorisation approach performed in our group has revealed that the endopeptidases hydrolysis of press cake generated peptide-phenol fractions with improved antioxidant, anti-inflammatory and anti-wrinkle activities important for anti-ageing properties, thereby setting these hydrolysates as potential ingredients for skin care applications [32].

LIMITATIONS AND POTENTIAL CHALLENGES

While the potential use of rapeseed co-streams into added-value applications in food, cosmetics and biomedical industry is an exciting and growing area of research, there are a few risks and limitations that should be tackled before the widespread use of rapeseed press cake/meal-derived protein products in human consumption. Hence, it will take time to realise their potential as edible and safe raw materials for cosmetics or nutraceutical industries.

Being novel food ingredients, rapeseed protein products will require special regulatory approval for marketing. To achieve this, trials are required that demonstrate the safety (animal/human testing data) of these compounds. Scientific data were presented to a regulatory agency, such as the US Food and Drug Administration (FDA) in order to confer a 'Generally Regarded As Safe' or 'GRAS' status for new food ingredients or processing aids. In the European Union and Canada, foods from non-traditional sources require approval under the Novel Food Legislation to ensure safety and clarity to the final consumer [60]. The first commercial rapeseed protein products by Burcon NutraScience

Corporation received GRAS status in 2008, and in 2014 *B. napus* protein isolates were approved for marketing in Europe as novel food ingredients.

Due to the absence of solid *in vivo* evidence and/or clinical data, frequency and administration of rapeseed bioactive compounds may be impossible to determine, leading to wide variability in intake and biological effects. Therefore, more investigation will be needed before they could be released to the market for human care [61]. Moreover, potential risk results from the presence of immunogenic proteins and peptides within the protein hydrolysates, which may induce and/or exacerbate allergic reactions in a minority of users [62, 63]. Another weak point is that due to the proteinaceous composition, rapeseed protein or peptide fractions may promote microbial contamination of skin care formulations. Therefore, the use of preservatives cannot be avoided.

Further technological advancements are necessary to generate economically feasible and sustainable production processes for rapeseed protein and bioactives. Commercialization of rapeseed protein isolates is currently challenged by the high investment and operational costs of water and energy-intensive wet fractionation processes [33]. Therefore, future research should focus on reducing water and chemical consumption to obtain ingredients with sufficient quality at reasonable costs [36]. Feasible recovery of high-quality fractions is challenged by the complex composition of rapeseed press cake/meal. Distinct properties of the two major rapeseed proteins open up wide application potential, but at the same time complicate their efficient recovery without multiple processing steps. Secondly, phytates and phenolic compounds in rapeseed press cake have a strong tendency to interact with rapeseed proteins and peptides and, therefore, are difficult to separate from the final products [31, 64]. Intensive research has been carried out and is still needed to prevent the co-fractionation of these compounds with rapeseed proteins and peptides.

Conclusion

Given their large quantities, the management of rapeseed oil industry co-streams is a growing focus of interest. In fact, over the past few decades industrial actors from oilseed producers to rapeseed crushers, showed interest in diversifying the markets for oilseed cake/meal, and in particular, to obtain high added-value products such as nutraceuticals, cosmeceuticals, etc. While many researchers use different technologies to isolate proteins, phenolics and other interesting biocompounds mainly for food applications, others are more focused on the valorisation of these co-streams as a whole, avoiding expensive purification steps.

The exploitation of these high protein industrial co-streams by using environmentally benign biomechanical and enzymatic processing technologies will surely bring new perspectives in food and skin care industries. In fact, the rapeseed industry, like many others that produce valuable waste and co-streams, offer plenty of potential and opportunities that may attract the attention of the stakeholders. Nevertheless, the adoption of ideas relevant to the biorefinery where rapeseed industry identifies these co-streams as potential sources of wealth aiming complete recycling and utilization are needed, rather than their rejection or application in low-profitable applications. Significant technological challenges need to be met before such changes are adopted by industry to satisfy current and future societal needs.

References

[1] Fediol. Composition and Quality of Vegetable Oils and Fats. *Rapeseed Oil*; 2011.

[2] FAO. *FAOSTAT Database* http://faostat3.fao.org (accessed Jun. 26, 2015).

[3] Kolesárová, N., Hutan, M., Bodík, I., Špalková, V. Utilization of Biodiesel by-Products for Biogas Production. *J. Biomed. Biotechnol.* 2011, 2011.

[4] Carré, P., Pouzet, A. Rapeseed Market, Worldwide and in Europe. *OCL* 2014, 21 (1), D102.

[5] Fediol. *Meals - EU Production, Imports, Exports and Consumption.*

[6] Leming, R., Lember, A. Chemical Composition of Expeller-Extraced and Cold-Pressed Rapeseed Cake. *Agraart. J. Agric. Sci.* 2005, 16 (2), 103-109.

[7] Eriksson, I., Westerlund, E., Åman, P. Chemical Composition in Varieties of Rapeseed and Turnip Rapeseed, Including Several Samples of Hull and Dehulled Seed. *J. Sci. Food Agric.* 1994, 66, 233-240.

[8] Bellostas, N., Sørensen, H., Sørensen, S. Nutrition and Processing Workshop: Quality of Rapeseed Oil for Non-Food (bioenergy), and Human and Animal Nutrition. In: *12th International Rapeseed Congress*; 2007; pp. 1-7.

[9] Ramachandran, S., Kumar, S., Larroche, C., Ricardo, C., Pandey, A. Oil Cakes and Their Biotechnological Applications – A Review. *Bioresour. Technol.* 2009, 98 (2007), 2000-2009.

[10] Kim, S. M., Na, M. S. A Study on Skin Care Effects of Rapeseed Meal Extract. *Korean Soc. Biotechnol. Bioeng.* 2013, 28 (3), 177-184.

[11] Pedroche, J., Yust, M. M., Megías, C., Lqari, H., Alaiz, M., Girón-calle, J., Millán, F., Vioque, J., Pedroche, P. J. Utilisation of Rapeseed Protein Isolates for Production of Peptides with Angiotensin I-Converting Enzyme (ACE) - Inhibitory Activity. *Grasas y aceites* 2004, 55, 354-358.

[12] Vioque, J., Sánchez-vioque, R., Clemente, A., Pedroche, J., Millán, F. Partially Hydrolyzed Rapeseed Protein Isolates with Improved Functional Properties. *J. Am. Oil Chem. Soc.* 2000, 77 (4), 447-450.

[13] McCarthy, A., O'Callaghan, Y., O'Brien, N. Protein Hydrolysates from Agricultural Crops-Bioactivity and Potential for Functional Food Development. *Agriculture* 2013, 3 (1), 112-130.

[14] Yoshie-Stark, Y., Wada, Y., Wäsche, A. Chemical Composition, Functional Properties, and Bioactivities of Rapeseed Protein Isolates. *Food Chem.* 2008, 107, 32-39.

[15] Schöne, F., Kirchheim, U., Schumann, W., Lüdke, H. Apparent Digestibility of High-Fat Rapeseed Press Cake in Growing Pigs and Effects on Feed Intake, Growth and Weight of Thyroid and Liver. *Anim. Feed Sci. Technol.* 1996, 62 (2), 97-110.

[16] Rommi, K., Hakala, T. K., Holopainen, U., Nordlund, E.; Poutanen, K., Lantto, R. Effect of Enzyme-Aided Cell Wall Disintegration on Protein Extractability from Intact and Dehulled Rapeseed (Brassica Rapa L. and Brassica Napus L.) Press Cakes. *J. Agric. Food Chem.* 2014, 62 (32), 7989-7997.

[17] Tenore, G. C., Troisi, J., Di Fiore, R., Basile, A., Novellino, E. Chemical Composition, Antioxidant and Antimicrobial Properties of Rapa Catozza Napoletana (Brassica Rapa L. Var. Rapa DC.) Seed Meal, a Promising Protein Source of Campania Region (southern Italy) Horticultural Germplasm. *J. Sci. Food Agric.* 2012, 92 (8), 1716-1724.

[18] WHO. Protein and Amino Acid Requirements in Human Nutrition; Geneva, 2007.

[19] Bos, C., Airinei, G., Mariotti, F., Benamouzig, R., Bérot, S.,Evrard, J., Fénart, E., Tomé, D., Gaudichon, C. The Poor Digestibility of Rapeseed Protein Is Balanced by Its Very High Metabolic Utilization in Humans. *J. Nutr.* 2007, 137 (3), 594-600.

[20] Wanasundara, J. P. D. Proteins of Brassicaceae Oilseeds and Their Potential as a Plant Protein Source. *Crit. Rev. Food Sci. Nutr.* 2011, 51 (7), 635-677.

[21] Aachary, A. A., Thiyam, U. A Pursuit of the Functional Nutritional and Bioactive Properties of Canola Proteins and Peptides. *Crit. Rev. Food Sci. Nutr.* 2012, 52 (11), 965-979.

[22] Kumar, V., Sinha, A. K., Makkar, H. P. S., Becker, K. Dietary Roles of Phytate and Phytase in Human Nutrition: A Review. *Food Chem.* 2010, 120 (4), 945-959.

[23] Cartea, M. E., Velasco, P. Glucosinolates in Brassica Foods: Bioavailability in Food and Significance for Human Health. *Phytochem. Rev.* 2008, 7 (2), 213-229.

[24] Bérot, S., Compoint, J. P., Larré, C., Malabat, C., Guéguen, J. Large Scale Purification of Rapeseed Proteins (Brassica Napus L.). *J. Chromatogr. B* 2005, 818 (1), 35-42.

[25] Ghodsvali, A., Khodaparast, M. H. H., Vosoughi, M., Diosady, L. L. Preparation of Canola Protein Materials Using Membrane Technology and Evaluation of Meals Functional Properties. *Food Res. Int.* 2005, 38 (2), 223-231.

[26] Xu, L., Diosady, L. L. Processing of Canola Proteins. A Review. In: *Canola and Rapeseed: Production, Processing, Food Quality, and Nutrition*; Thiyam-Holländer, U., Eskin, M. N. A., Matthäus, B., Eds.; CRC Press: Boca Raton, 2012; pp. 59-78.

[27] Sari, Y. W., Bruins, M. E., Sanders, J. P. M. Enzyme Assisted Protein Extraction from Rapeseed, Soybean, and Microalgae Meals. *Ind. Crops Prod.* 2013, 43 (1), 78-83.

[28] Niu, Y. X., Li, W., Zhu, J., Huang, Q., Jiang, M., Huang, F. Aqueous Enzymatic Extraction of Rapeseed Oil and Protein from Dehulled Cold-Pressed Double-Low Rapeseed Cake. *Int. J. Food Eng.* 2012, 8 (3), 1-14.

[29] DeCastro, F. B., Carlsson, T. I., Kvist, S. U., Lawther, J. M. *Process for the Fractionation of Oilseed Press Cakes and Meals*, 2005.

[30] Ma, C.-Y., Harwalkar, V. R., Paquet, A. Physicochemical Properties of Alkali-Treated Oat Globulin. *J. Agric. Food Chem.* 1990, 38, 1707-1711.

[31] Xu, L., Diosady, L. L. Interactions between Canola Proteins and Phenolic Compounds in Aqueous Media. *Food Res. Int.* 2000, 33 (9), 725-731.

[32] Rivera, D., Rommi, K., Fernandes, M. M., Lantto, R., Tzanov, T. Biocompounds from Rapeseed Oil Industry Co-Stream as Active Ingredients for Skin Care Applications. *Int. J. Cosmet. Sci.* 2015, 37, 496-505.

[33] Schutyser, M. A. I., van der Goot, A. J. The Potential of Dry Fractionation Processes for Sustainable Plant Protein Production. *Trends Food Sci. Technol.* 2011, 22 (4), 154-164.

[34] Rommi, K., Holopainen, U., Pohjola, S., Hakala, T. K., Lantto, R., Poutanen, K., Nordlund, E. Impact of Particle Size Reduction and Carbohydrate-Hydrolyzing Enzyme Treatment on Protein Recovery from Rapeseed (Brassica Rapa L.) Press Cake. *Food Bioprocess Technol.* 2015.

[35] Rommi, K., Ercili-Cura, D., Hakala, T. K., Nordlund, E., Poutanen, K., Lantto, R. Impact of Total Solid Content and Extraction pH on Enzyme-Aided Recovery of Protein from Defatted Rapeseed (Brassica Rapa L.) Press Cake and Physicochemical Properties of the Protein Fractions. *J. Agric. Food Chem.* 2015, 63 (11), 2997-3003.

[36] Von Der Haar, D., Müller, K., Bader-Mittermaier, S., Eisner, P. Rapeseed Proteins – Production Methods and Possible Application Ranges. *Oilseeds fats Crop. Lipids* 2014, 21 (1), 1-8.

[37] Hu, Z.-Y., Hua, W., Zhang, L., Deng, L.-B., Wang, X.-F., Liu, G.-H., Hao, W.-J., Wang, H.-Z. Seed Structure Characteristics to Form Ultrahigh Oil Content in Rapeseed. *PLoS One* 2013, 8 (4), 1-10.

[38] Naczk, M., Amarowicz, R., Shahidi, F. Role of Phenolics in Flavor of Rapeseed Protein Products. *Dev. Food Sci.* 1998, 40, 597-613.

[39] Pustjens, A. M., Schols, H. A., Kabel, M. A., Gruppen, H. Characterisation of Cell Wall Polysaccharides from Rapeseed (Brassica Napus) Meal. *Carbohydr. Polym.* 2013, 98, 1650-1656.

[40] Akbari, A., Wu, J. An Integrated Method of Isolating Napin and Cruciferin from Defatted Canola Meal. *LWT - Food Sci. Technol. Sci. Technol.* 2015, 64, 308-315.

[41] Zhou, B., He, Z., Yu, H., Mukherjee, K. D. Proteins from Double-Zero Rapeseed. *J. Agric. Food Chem.* 1990, 38 (3), 690-694.

[42] Sari, Y. W., Syafitri, U., Sanders, J. P. M., Bruins, M. E. How Biomass Composition Determines Protein Extractability. *Ind. Crops Prod.* 2015, 70, 125-133.

[43] Klassen, D. R., Elmer, C. M., Nickerson, M. T. Associative Phase Separation Involving Canola Protein Isolate with Both Sulphated and Carboxylated Polysaccharides. *Food Chem.* 2011, 126 (3), 1094-1101.

[44] Stone, A. K., Teymurova, A., Dang, Q., Abeysekara, S., Karalash, A., Nickerson, M. T. Formation and Functional Attributes of Electrostatic Complexes Involving Napin Protein Isolate and Anionic Polysaccharides. *Eur. Food Res. Technol.* 2014, 238 (5), 773-780.

[45] Vioque, J., Sanchez-Vioque, R., Clemente, A., Pedroche, J., Bautista, J., Millan, F. Production and Characterization of an Extensive Rapeseed Protein Hydrolysate. *J. Am. Oil Chem. Soc.* 1999, 76 (7), 819-823.

[46] Kim, H.-O., Li-Chan, E. C. Y. Quantitative Structure-Activity Relationship Study of Bitter Peptides. *J. Agric. Food Chem.* 2006, 54 (26), 10102-10111.

[47] Wu, J., Aluko, R. E., Muir, A. D. Purification of Angiotensin I-Converting Enzyme-Inhibitory Peptides from the Enzymatic Hydrolysate of Defatted Canola Meal. *Food Chem.* 2008, 111 (4), 942-950.

[48] Yoshie-Stark, Y., Wada, Y., Schott, M., Wäsche, A. Functional and Bioactive Properties of Rapeseed Protein Concentrates and Sensory Analysis of Food Application with Rapeseed Protein Concentrates. *LWT - Food Sci. Technol.* 2006, 39 (5), 503-512.

[49] Pudel, F. Recent Results on Extraction and Application. In: Canola *and Rapeseed Production, Processing, Food Quality, and Nutrition*; Thiyam-Holländer, U., Michael Eskin, N. A., Matthäus, B., Eds.; CRC Press, Taylor and Francis Group, 2012; pp. 186-202.

[50] Lomascolo, A., Uzan-Boukhris, E., Sigoillot, J.-C., Fine, F. Rapeseed and Sunflower Meal: A Review on Biotechnology Status and Challenges. *Appl. Microbiol. Biotechnol.* 2012, 95 (5), 1105-1114.

[51] Frost and Sullivan. Analysis of the Global Protein Ingredients Market; 2013.

[52] Stensson Anjou, K. O., Jeny Fecske, A., Car, G. K., Ragnar Ohlson, J. S. Production of Rapeseed Protein Concentrate for Human Consumption, 1978.

[53] Krause, J.-P., Schwenke, K. D. Behaviour of a Protein Isolate from Rapeseed (Brassica Napus) and Its Main Protein Components - Globulin and Albumin - at Air/solution and Solid Interfaces, and in Emulsions. *Colloids Surf. B. Biointerfaces* 2001, 21 (1-3), 29-36.

[54] Rubino, M. I., Amfield, S. D., Nadon, C. A., Bernatsky, A. Phenolic Protein Interactions in Relation to the Gelation Properties of Canola Protein. *Food Res. Int.* 1996, 29 (7), 653-659.

[55] Partanen, R., Sibakov, J., Rommi, K., Hakala, T., Holopainen-Mantila, U., Lahtinen, P., Ercili Cura, D., Lantto, R. Modifying the Colloidal

State of Non-Refined Rapeseed (Brassica Rapa) Protein Concentrate by Microfluidization.

[56] Yoshie-Stark, Y., Wada, Y., Schott, M., Wäsche, A. Functional and Bioactive Properties of Rapeseed Protein Concentrates and Sensory Analysis of Food Application with Rapeseed Protein Concentrates. *LWT - Food Sci. Technol.* 2006, 39 (5), 503-512.

[57] Chen, Y., Thiyam-Hollander, U., Barthet, V. J., Aachary, A. A. Value-Added Potential of Expeller-Pressed Canola Oil Refining: Characterization of Sinapic Acid Derivatives and Tocopherols from Byproducts. *J. Agric. Food Chem.* 2014, 62 (40), 9800-9807.

[58] Sørensen, A.-D. M. D. M., Friel, J., Winkler-Moser, J. K., Jacobsen, C., Huidrom, D., Reddy, N., Thiyam-Holländer, U. Impact of Endogenous Canola Phenolics on the Oxidative Stability of Oil-in-Water Emulsions. *Eur. J. Lipid Sci. Technol.* 2013, 115 (5), 501-512.

[59] Thiyam, U., Stöckmann, H., Zum Felde, T., Schwarz, K. Antioxidative Effect of the Main Sinapic Acid Derivatives from Rapeseed and Mustard Oil by-Products. *Eur. J. Lipid Sci. Technol.* 2006, 108 (3), 239-248.

[60] Arntfield, S. D. Canola and Other Oilseed Proteins. In: *Handbook of food proteins*; Woodhead Publishing Limited: Manitoba, 2011; pp. 289-315.

[61] Chakrabarti, S., Forough, J., Wu, J., Jahandideh, F., Wu, J. Food-Derived Bioactive Peptides on Inflammation and Oxidative Stress. *Biomed Res. Int.* 2014, 2014, DOI: 10.1155/2014/608979.

[62] Akiyama, H., Sakata, K., Yoshioka, Y., Murata, Y., Ishihara, Y., Teshima, R., Sawada, J., Maitani, T. Profile Analysis and Immunoglobulin E Reactivity of Wheat Protein Hydrolysates. *Int. Arch. Allergy Immunol.* 2006, 140 (1), 36-42.

[63] Puumalainen, T. J., Puustinen, A., Poikonen, S., Turjanmaa, K., Palosuo, T., Vaali, K. Proteomic Identification of Allergenic Seed Proteins, Napin and Cruciferin, from Cold-Pressed Rapeseed Oils. *Food Chem.* 2015, 175, 381-385.

[64] Schwenke, K. D., Mothes, R., Marzilger, K., Borowska, J., Kozlowska, H. Rapeseed Protein Polyanion Interactions - Turbidimetric Studies in Systems with Phosphate-Containing Polyanions: Phytic Acid and Octametaphosphate. *Food/Nahrung* 1987, 31 (10), 1001-1013.

In: Rapeseed
Editor: Monica White

ISBN: 978-1-63484-227-3

Chapter 3

INFLUENCE OF SUNFLOWER AND RAPESEED LECITHIN ON PHYSICAL PROPERTIES OF FAT FILLING

Ivana Lončarević*[*], *Biljana Pajin and Jovana Petrović
Faculty of Technology, University of Novi Sad, Novi Sad, Serbia

ABSTRACT

Vegetable lecithins are commercially obtained from oil-bearing seeds such as soybeans, sunflower and rapeseed. Soybean seeds are the main source of lecithin in confectionery industry in Serbia and elsewhere. On the other hand, production of sunflower and rapeseed oil makes economically sustainable production of sunflower and rapeseed lecithin and their use as emulsifiers in different confectionery products.

This chapter examined physical properties of fat filling produced with oil industry by-products – sunflower and rapeseed lecithin in comparison with widely used emulsifier - soy lecithin. Rheological, textural and thermal properties were determined by rotational viscometry, texture analyser, and differential scanning calorimetry while particle size distribution was analysed using particle size analyser.

The obtained results showed similar particle size distribution of fat filling produced with soy, sunflower and rapeseed lecithin within each applied milling time, indicating appropriate production. Regarding rheological properties, fat filling samples with sunflower lecithin had lower values of viscosity and textural parameters compared to samples with soy and rapeseed lecithin at all applied milling time. On the other hand, soy lecithin influenced higher values of mentioned parameters. Prolonging the milling time had no significant influence on termal parameters of fat filling produced with all three type of lecithin where samples with sunflower lecithin began to melt at lower temperatures in relation to samples with soy and rapeseed lecithin.

INTRODUCTION

During oil processing, the phospho- and glycolipids present in the crude oil are removed in order to stabilize vegetable oil against sedimentation and also to enable further refining steps. The obtained crude lecithin, as a by-product of vegetable oil refining process, can be defined as a mixture of polar lipids (phospholipids, glycolipids), carbohydrates and vegetable oil together with minor components [1]. The often used quality characteristic acetone-insoluble matter includes phospholipids, glycolipids and carbohydrates all together [2].

Soybean oil has been the primary source of commercial vegetable lecithin for many years, typically containing between 0.5 and 3% of phospholipids [3]. Only a part of the potential lecithin gums of the crushed seeds is processed as lecithin. The world market for commercial lecithin is estimated at 150,000-170,000 metric tons, making it a leading food emulsifier group. Sourcing of identity-preserved (IP) non-GMO soy lecithin for the European market may not meet all these demands, which presents a market opportunity for high-quality IP sunflower and rapeseed lecithins [4]. Although the percentage of phospholipids in sunflower oil averages around 0.75%, the availability of lecithin from this oil may be a possibility assuming the increased production of sunflower oil in the last years. Sunflower lecithin has a mild taste with similar phospholipid composition to soybean lecithin, with a tendency to higher PC and lower PE ratios, which might be caused by crop varieties and processing conditions [5]. On the other hand, Rapeseed (*Brassica napus L.*) is the second most important oilseed crop in the world after soybean [6]. At first rapeseed lecithin was applied as an emulsifier and energy component in animal feed. Today, as double-zero rapeseed varieties, such as canola, became available the applications of rapeseed lecithin have developed positively [5].

Functional properties of lecithin are mainly caused by the surface-active character of its phospholipids. These consist of a glycerol backbone esterified with two fatty acids and a phosphate group which may be esterified with monovalent alcohols (for example choline or ethanolamine), or polyvalent alcohols (for example glycerol or inositol). The phospholipid and fatty acid composition of lecithin usually depend on type of oil from which lecithin was obtained and on processing conditions [7]. Phosphatidylcholines (PC), phosphatidylethanolamines (PE), phosphatidylinositols (PI), phosphatidylserines (PS), phosphatidilglycerols (PG), phosphatidic acid (PA), glycolphospholipids (GPL) and their lysoforms are the major phospholipids found in lecithin from vegetable sources [8].

Lecithins as food additives are legally regulated in the European Union under the number E 322. In addition to definition and appearance the quality requirements for lecithin also include: not less than 60% acetone insoluble substances; less than 0.3% toluene insoluble substances; less than 2% volatile substances (determined after 1 h drying at 105°C); less than 35 mg potassium hydroxide per gram; peroxide value 10 or less, expressed in mequ/kg [1]. The amount of acetone insoluble matter (%AI) is the approximate indication for the amount of phospholipids, glycolipids and carbohydrates, since the oil and fatty acids dissolve in acetone. On the other hand, the level of toluene insoluble matter (TI) is a measure of the purity of lecithins. TI matter usually consists of residual fiber, but sometimes particulate contaminants may be introduced during processing [2].

Application of Lecithin in Fat Based Confectionery Products

A lot of confectionery products contain a substantial amount of fat phase which contributes to the structure of the final product. Many of the sensory attributes of fat based confectionery products, such is spreadability, mouth feel, texture, etc. are dependent on the macroscopic properties of the fat crystal network [9, 10]. As a continuous phase in fat based confectionery products the fat phase completely determines the hardening and melting as well as the consistency of the final product. Therefore, the fat selection depends on its characteristics and complex processes that may occur during manufacture and later in storage [11]. Beside carefully selected raw materials, industrially produced food requires emulsifiers to facilitate processing and ensure final product with a uniform quality and long shelf-life [4]. As amphipathic molecules lecithins have found numerous applications in the food industry, mainly as emulsifiers and stabilizers [8]. Lecithin helps to smooth the texture of food and serves as an emulsifying agent in the manufacture of chocolate, bakery and instant products, margarines, and mayonnaise [12, 13]. One of the most traditional applications for lecithin is its use in chocolate production [14]. Chocolate mass is a suspension of solid sugar, milk powder and cocoa particles dispersed in a continuous cocoa butter phase [15]. At the end of the conch process the all amount of lecithin (0.3-0.5%) is usually added in all amount. Lecithin has effect on chocolate mass flow properties in terms of viscosity reduction, flow improvement during tempering, and flow improvement during molding or enrobing. The flow properties of chocolate mass is measured at 40°C since the fat is molten with no fat crystals present [2]. Thus, chocolate rheology is mainly defined by fat content, applied emulsifier, processing conditions and solid particle size distribution [16]. The main difference between chocolate and other cocoa based products is in fat phase. For example, cocoa spreads does not contain cocoa butter but contains cheaper vegetable fats, mostly hydrogenated oils, and may contain vegetable oil to improve spreadability [17]. The main characteristics of this type of product are: good spreadability in wide temperature range (from ambient to fridge), rich creamy taste and homogeneous smooth structure, without fat phase migration on the cream surface [18].

Being a surface active phospholipid mixture, lecithin performs as an excellent wetting and emulsifying agent. In heterogeneous systems the phospholipid molecules arrange themselves in a monomolecular layer with the fatty acid facing the oil surface and the phosphoric acid portion facing the water surface. The arrangement lowers the interfacial tension of the oil-water boundary with resultant benefits such as rapid wetting, lowering of viscosity, and better and more stable emulsions or dispersions [19]. In manufacture of liquid chocolate mass and similar oil-based suspensions lecithin is commonly used as surfactant. The addition of lecithin to oil-based suspensions causes adsorbing of surface active components at the surface of suspended particles, and thus smoothing of particle surface. This reduces the friction between particles and results in a decrease of yield stress and viscosity until a surfactant concentration for maximum reduction is reached. Further increasing lecithin concentration raises yield stress and does not lead to a further reduction of viscosity [7]. Lecithin is added in relatively small amounts (0.1–2%) as emulsifier in food production whereby its color, odor and flavor are not normally noticeable [20].

This chapter investigated physical properties of fat filling produced in a laboratory ball mill applying standard soy lecithin as emulsifier, as well as sunflower and rapeseed lecithin, analyzing their influence on physical properties of fat filling under three applied milling times.

MATERIALS AND METHODS

Starting Materials

- Fat filling mass that passed through 3 roll mill in industrial conditions - mixture of powdered sugar ("Crvenka a.d.," Crvenka, Serbia), cocoa powder ("Centroproizvod," Beograd, Serbia), soy flour ("Sojaprotein," Bečej, Serbia), milk powder ("Imlek," Beograd, Serbia), vegetable fat ("Dijamant," Zrenjanjin, Serbia)
- K-VEGETABLE fat (Dijamant, Zrenjanin, Serbia). Fat characteristics - fatty acid composition and solid fat content at different temperatures are given in previous research [9].
- Native fluid soybean lecithin, containing more than 60% matter insoluble in acethone, and less than 0.3% matter insoluble in toluene, produced and declared in "Victoriaoil" Oil Factory, Šid, Serbia
- Native fluid sunflower lecithin, containing more than 60% matter insoluble in acethone, and less than 0.3% matter insoluble in toluene, produced and declared in "Victoriaoil" Oil Factory, Šid, Serbia
- Native fluid rapeseed lecithin, containing more than 60% matter insoluble in acethone, and less than 0.3% matter insoluble in toluene, produced and declared in "Victoriaoil" Oil Factory, Šid, Serbia

The composition of fat filling includes: 50% of powdered sugar, 30% of vegetable fat, 7% of cocoa powder, 5.5% of soy flour, 7% of milk powder, 0.5% of lecithin.

Plan of Experiments

Fat filling was produced in laboratory ball mil, with capacity of 5 kg. The samples were prepared with 0.5% of soy, sunflower and rapeseed lecithin, with different time of milling (30, 50, and 70 minutes) in order to get the following samples:

Lecithin	Soy			Sunflower			Rapeseed		
Milling time (min)	30	50	70	30	50	70	30	50	70
Sample	Soy 30	Soy 50	Soy 70	Sun 30	Sun 50	Sun 70	Rape 30	Rape 50	Rape 70

Methods

Fat filling production. Fat and lecithin were homogenized first in laboratory ball mill for 5 minutes. Then fat filling mass was added. The temperature in the ball mill was 35°C, with a rotation speed of 50 rpm. Fat filling samples were added into a sterile plastic cups after defined milling time and each cup capped with a plastic lid.

Particle size distribution. Particle size distribution in fat filling samples was determined by Mastersizer laser diffraction particle size analyzer equipped with Hydro 2000 μP wet dispersion unit (Malvern Instruments). Fat filling sample was dispersed in sunflower oil at ambient temperature (21 ± 1°C) and added until adequate obscuration was obtained. All measurements were performed in triplicate. Obtained particle size distribution parameters included mass median diameter d(0.5); parameters d(0.1) and d(0.9), and volume weighted mean d_{mm}.

Rheological properties. Rheological properties of fat filling samples were determined by rotational rheometer Rheo Stress 600, Haake. The tests were carried out at 35°C using a concentric cylinder system (Z20 DIN). The shear rate was increased from 0 s^{-1} to 60 s^{-1} within a period time of 180 s, then was kept constant at max. speed of 60 s^{-1} 60 s and after that was reduced from 60 s^{-1} to 0 s^{-1}, within 180 s, according to modified O.I.C.C.C. method [21].

Textural characteristics. Textural characteristics of fat samples were analyzed using a Texture Analyzer TA.XT Plus, Stable Micro System. The hardness and work of shearing were determined by penetration at ambient temperature of 21 ± 1°C, according to method Chocolate Spread – SPRD2_SR_PRJ. Each sample was placed into the cone sample holder and pressed down in order to eliminate air pockets. Any excess of sample was scraped off with a knife. Then the filled cone sample holder was put in base holder and 45 degree cone probe was used to penetrate the samples at 3 mm/s.

Thermal properties. Differential scanning calorimetry DSC 910, Termal analyzer 990 and Dynamic mechanical analyzer (Du Point Instruments, USA) was used to determine the thermal profile of fat filling samples produced with soy, sunflower and rapeseed lecithin. 5 mg of fat filling samples were weighed in aluminium pans and pierced covers were sealed in place. An empty, hermetically sealed aluminium pan was used as a reference. Samples were analyzed by the following procedure: they were heated from 10°C to 100°C with heating rate of 5°C per minute.

Statistical analysis. Results were expressed as mean of triplicate analyses. The results were statistically tested using ANOVA method and the means were compared by one-factor analysis at variance with subsequent comparisons by Duncan's test at a significance level at 0.05 using software Statistica 12.0 (Statsoft, USA).

Results and Discussion

Particle Size Distribution

Particle size distribution of fat filling produced with lecithin from different origin, milled 30, 50, and 70 minutes in laboratory ball mill is presented in Figure 1.

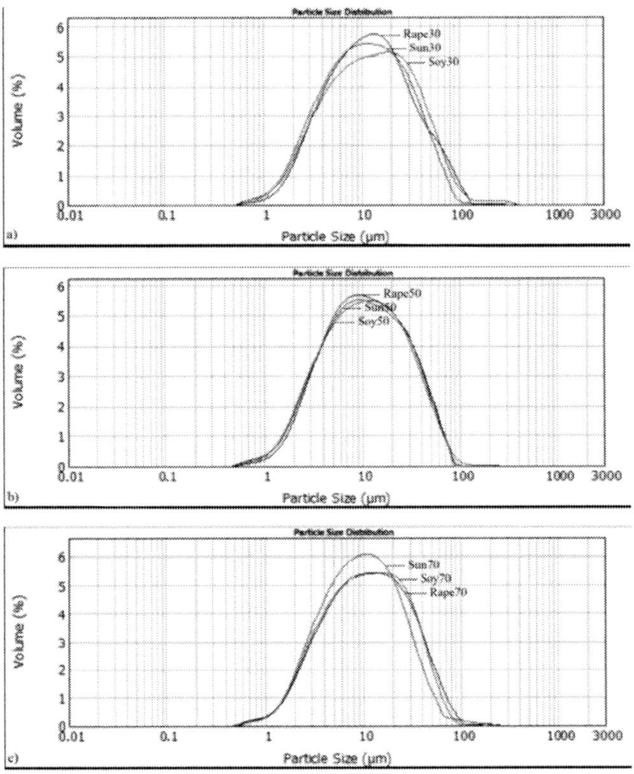

Figure 1. Particle size distribution of fat filling samples with soy, sunflower and rapeseed lecithin, depending on milling time: a) 30; b) 50; and c) 70 minutes in a ball mill.

Obtained curves represent the Gaussian distribution of solid particles and relatively uniform particle sizes in fat fillings with soy, sunflower and rapeseed lecithin, depending on each applied milling time in laboratory ball mill. However, particle size results in terms of d(0.1), d(0.5), d(0.9) and d_{mm} show differences between the samples and are presented in Table 1.

Table 1. Parameters of particle size distribution of fat filling

Sample	Particle size distribution (μm)			d_{mm}
	d(0.1)	d(0.5)	d(0.9)	
Soy30	$3.06^{b} \pm 0.17$	$12.75^{e} \pm 0.33$	$46.87^{c} \pm 3.21$	$20.98^{d} \pm 2.85$
Soy50	$2.78^{a} \pm 0.04$	$10.55^{b} \pm 0.27$	$36.73^{b} \pm 0.86$	$16.06^{ab} \pm 0.50$
Soy70	$2.84^{a} \pm 0.13$	$10.82^{bc} \pm 0.21$	$35.89^{b} \pm 0.83$	$15.73^{ab} \pm 0.71$
Sun30	$2.90^{a} \pm 0.06$	$11.01^{c} \pm 0.08$	$38.80^{b} \pm 1.30$	$17.88^{c} \pm 0.28$
Sun50	$2.87^{a} \pm 0.16$	$10.56^{b} \pm 0.17$	$37.47^{b} \pm 0.50$	$15.92^{ab} \pm 0.20$
Sun70	$2.81^{a} \pm 0.04$	$9.59^{a} \pm 0.17$	$30.24^{a} \pm 1.11$	$14.08^{a} \pm 0.63$
Rape30	$3.27^{c} \pm 0.06$	$11.92^{d} \pm 0.08$	$44.93^{c} \pm 0.58$	$18.91^{cd} \pm 0.22$
Rape50	$3.08^{b} \pm 0.12$	$11.13^{c} \pm 0.39$	$38.69^{b} \pm 4.34$	$16.89^{bc} \pm 2.06$
Rape70	$2.92^{b} \pm 0.05$	$10.72^{bc} \pm 0.16$	$36.94^{b} \pm 0.85$	$15.98^{ab} \pm 0.27$

Values represent the means of three measurements ± standard deviation. Values followed by different lower-case letters in the same column are significantly different from each other ($p < 0.05$).

Prolonging the milling time generally decreases all particle size parameters within fat filling samples with particular applied lecithin, where samples with rapeseed lecithin have significantly ($p < 0.05$) higher values of d(0.1) and d(0.5) in relation to samples with soy and sunflower lecithin within all applied milling time (with the exception of d(0.5) for Rape70). On the other hand, there are no significantly ($p < 0.05$) different values of d(0.9) between fat filling samples produced with soy and rapeseed lecithin within all applied milling times, while Sun30 and Sun70 have statistically significant ($p < 0.05$) lower values of d(0.9).

Despite significant differences between some particle size parameters in fat fillings within each applied milling time, the values of volume weighted mean d_{mm} do not significantly differs ($p < 0.05$). The only exception is Soy30 which has statistically significant higher value of d_{mm} in comparison with fat filling sample produced with sunflower lecithin, milled 30 minutes in laboratory ball mill.

Rheological Properties

Figure 2 presents flow curves of fat filling samples, determined at temperature of fat filling production (35°C).

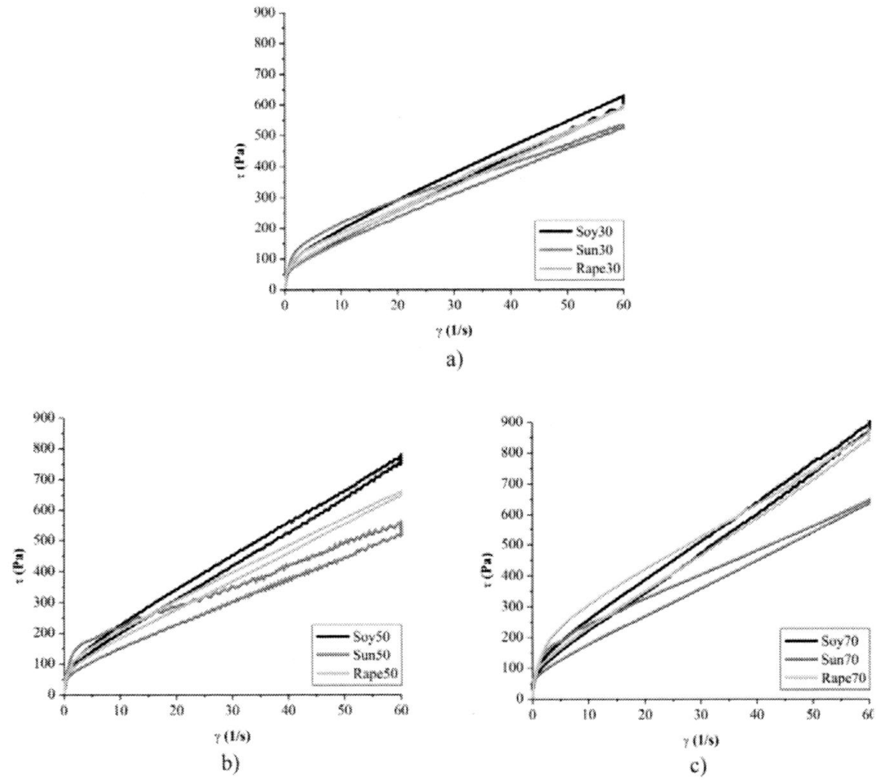

Figure 2. Flow curves of fat filling samples with soy, sunflower and rapeseed lecithin milled: a) 30, b) 50, and c) 70 minutes in a ball mill.

Since the values of volume weighted mean d_{mm} generally do not significantly differs ($p < 0.05$) in fat filling samples with soy, sunflower and rapeseed lecithin within all applied milling time, it can be assumed that rheological properties mostly depend on lecithin type.

All samples show thixotropic flow. It is evident that fat filling samples with soy lecithin generally have higher values of shear stress at higher shear rate, within particular milling time. On the other hand, samples with sunflower and rapeseed lecithin have higher values of shear stress at lower shear rate compared to fat filling samples produced with soy lecithin. Samples of fat filling with sunflower lecithin have lower values of viscosity at all applied milling time. Rheological parameters are given in Table 2.

Table 2. Rheological parameters of fat filling samples

Sample	Thixotropy curve area (Pa/s)	Casson yield stress (Pa)	Viscosity at maximum shear rate (Pas)	Casson viscosity (Pas)
Soy30	$2192.25^{c} \pm 22.89$	$39.82^{b} \pm 1.36$	$10.21^{c} \pm 0.05$	$5.82^{e} \pm 0.03$
Soy50	$2950.54^{e} \pm 27.60$	$37.75^{a} \pm 1.48$	$12.75^{f} \pm 0.06$	$7.70^{g} \pm 0.02$
Soy70	$3230.48^{f} \pm 30.21$	$40.77^{b} \pm 2.01$	$14.76^{h} \pm 0.08$	$9.02^{h} \pm 0.05$
Sun30	$2728.12^{d} \pm 25.63$	$75.58^{f} \pm 2.69$	$8.85^{a} \pm 0.05$	$3.40^{a} \pm 0.01$
Sun50	$7051.34^{h} \pm 36.98$	$73.67^{e} \pm 2.21$	$8.89^{a} \pm 0.03$	$3.61^{b} \pm 0.02$
Sun70	$2757.17^{d} \pm 28.69$	$72.31^{e} \pm 3.01$	$10.72^{d} \pm 0.04$	$4.61^{c} \pm 0.04$
Rape30	$974.09^{a} \pm 15.34$	$36.21^{a} \pm 1.98$	$9.94^{b} \pm 0.05$	$5.52^{d} \pm 0.03$
Rape50	$1819.46^{b} \pm 19.36$	$43.35^{c} \pm 1.64$	$10.93^{e} \pm 0.06$	$6.01^{e} \pm 0.03$
Rape70	$4189.61^{g} \pm 31.24$	$68.44^{d} \pm 2.96$	$14.25^{g} \pm 0.06$	$6.79^{f} \pm 0.02$

Values represent the means of three measurements ± standard deviation. Values followed by different lower-case letters in the same column are significantly different from each other ($p < 0.05$).

The data given in Table 2 shows that prolonging the milling time leads to increasing the viscosity at maximum shear rate and Casson viscosity in fat filling produced with all three type of lecithin. This is due to larger surface area of all particles in samples with longer milling time where the system is more compact with less free fat phase. Fat filling with sunflower lecithin have statistically significant ($p < 0.05$) higher values of Casson yield stress or higher initial force that must be applied to the system to begin to flow, compared to samples with soy and rapeseed lecithin. However, those samples have lower values of viscosity at maximum shear rate and Casson viscosity at particular milling time in relation to fat filling samples produced with soy and rapeseed lecithin. On the other hand, the highest values of viscosity are noticed in samples with soy lecithin, within all applied milling time.

Prolonging the milling time causes higher values of tixotropy curve area in samples with the addition of soy and rapeseed lecithin, mostly pronounced in Rape70. This means that increasing the shear rate in those samples influences more pronounced destruction of the structure caused by breaking the bonds within packed sugar crystals and cocoa solids [16]. On the other hand, fat filling sample with sunflower lecithin milled 50 minutes in ball mill have the highest value of tixotropy curve area, that significantly differs ($p < 0.05$) in relation to all other samples.

Textural Characteristics

Figure 3 shows textural characteristics of fat filling samples with the addition of lecithins from different origin, milled 30, 50 and 70 minutes in ball mill.

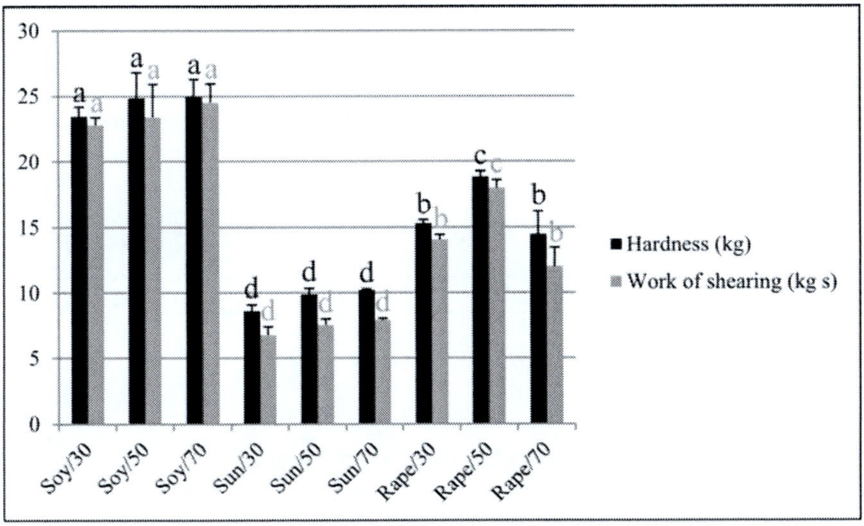

Values represent the means; n=3. Values followed by different lower-case letters (within hardness and work of shearing) are significantly different from each other ($p<0.05$).

Figure 3. Textural characteristics of fat filling samples with soy, sunflower and rapeseed lecithin depending on milling time.

Samples with soy lecithin have a significantly ($p < 0.05$) higher values of hardness and work of shearing at 20°C comparing to other fat filling samples with sunflower and rapeseed lecithin. On the other hand, samples with sunflower lecithin have the lowest statistically significant ($p < 0.05$) values of textural parameters at each milling time. The addition of rapeseed lecithin reduces the values of textural parameters in relation to samples with soy lecithin, but not as much as the addition of sunflower lecithin. The obtained results are in accordance with rheological measurements that showed the highest viscosity values in fat filling samples with soy lecithin and the lowest in sample with sunflower lecithin. The research [4] also showed that fat samples with addition of sunflower lecithin had the lowest values of textural parameters, due to the consistency of sunflower lecithin that is more fluid and less viscous than soybean and rapeseed lecithin. Prolonging the milling time slightly increases the values of textural parameters in all samples (with the exception in sample Rape/70) due to lower particle size of solids that cause more compact structure of fat filling. However, the results within each applied lecithin do not differ statistically significant ($p < 0.05$) with also the exception in sample with rapeseed lecithin milled 70 minutes in a ball mill.

Thermal Properties

Table 3 shows DSC parameters - onset temperature (T_o), peak temperature (T_P), and conclusion temperature (T_c).

Table 3. Thermal parameters of fat filling samples

Sample	T_{onset} (°C)	T_{end} (°C)	T_{peak} (°C)	T_{index} [T_{end} - T_{onset}]
Soy30	$31.14^{c} \pm 0.17$	$37.35^{ab} \pm 0.31$	$33.54^{cde} \pm 0.56$	6.21^{ab}
Soy50	$31.25^{c} \pm 0.27$	$37.89^{b} \pm 0.69$	$33.25^{bcd} \pm 0.29$	6.64^{ab}
Soy70	$30.82^{bc} \pm 0.35$	$37.42^{ab} \pm 0.24$	$32.92^{bc} \pm 0.35$	6.60^{ab}
Sun30	$29.35^{a} \pm 1.04$	$36.24^{a} \pm 0.68$	$34.35^{e} \pm 0.12$	6.89^{b}
Sun50	$30.17^{ab} \pm 0.34$	$36.68^{ab} \pm 0.90$	$34.12^{e} \pm 0.65$	6.51^{ab}
Sun70	$29.63^{a} \pm 0.69$	$36.52^{a} \pm 0.84$	$34.36^{e} \pm 0.26$	6.89^{b}
Rape30	$30.54^{bc} \pm 0.32$	$36.90^{ab} \pm 0.42$	$32.65^{abc} \pm 0.92$	6.36^{ab}
Rape50	$30.69^{bc} \pm 0.13$	$36.84^{ab} \pm 0.52$	$32.26^{ab} \pm 0.34$	6.15^{a}
Rape70	$30.22^{ab} \pm 0.26$	$36.74^{ab} \pm 0.68$	$31.89^{a} \pm 0.91$	6.52^{ab}

Values represent the means of three measurements ± standard deviation. Values followed by different lower-case letters in the same column are significantly different from each other ($p < 0.05$).

Melting first begins in fat filling samples with sunflower lecithin (29.35–30.17°C) while the samples with soy lecithin have higher values of T_{onset} in comparison to samples with sunflower and rapeseed lecithin, ranging in interval from 30.8 2°C in Soy70 to 31.25 in Soy50. Having higher temperatures of T_{onset} the samples with soy lecithin also have higher T_{end} temperatures (37.35–37.89°C) compared to fat filling samples produced with sunflower and rapeseed lecithin, which, on the other hand, are not significantly different ($p < 0.05$) in relation to T_{end} values of fat filling with sunflower and rapeseed lecithin. The addition of sunflower lecithin affect the melting piont of fat filling where the samples with sunflower lecithin have significantly higher values ($p < 0.05$) of T_{peak} in comparison to other samples. However, T_{index} range in interval from 6.15 in Rape50 to 6.89 in Sun30 and Sun70 with generally no statistically differences ($p < 0.05$) between all samples.

Generally, prolonging the milling time from 30 to 50 and 70 minutes has no significant influence ($p < 0.05$) on termal parameters of fat filling produced with all three type of lecithin.

CONCLUSION

The aim of this chapter was to compare physical characteristics of fat filling produced with standard soy lecithin with fat filling produced with sunflower and rapeseed lecithin, depending on milling time in laboratory ball mill. Obtained results showed that all applied milling time were enough for achieving adequate degree of fragmentation of solid particles in fat filling samples which should provide adequate melting in the mouth during consumption. However, prolonging the milling time reduced particle size in all samples which caused more compact structure and thus generally increased values of viscosity and textural parameters in all samples within particular lecithin, but with no influence on their termal properties. Samples with sunflower lecithin had higher values of yield stress but, on the other hand, lower values of viscosity and textural parameters within all applied milling time compared to samples with soy and rapeseed lecithin and also began to melt at lower temperatures.

Acknowledgments

This chapter has been supported by the Ministry of Science and Technological Development of the Republic of Serbia (Project no. 31014).

References

[1] Whitehurs, R. J. (2004). *Emulsifiers in Food Technology.* Oxford: Blackwell Publishing Ltd.

[2] Nieuwenhuyzen, W., & Tomas, M. C. (2008). Update on vegetable lecithin and phospholipid technologies. *European Journal of Lipid Science and Technology,* 110, 472-486.

[3] Doig, S. D., & Diks, R. M. M. (2003). Toolbox for exchanging constituent fatty acids in lecithins. *European Journal of Lipid Science and Technology,* 105, 359–367.

[4] Lončarević, I., Pajin, B., Omorjan, R., Torbica, A., Zarić, D., Maksimović, J., & Švarc Gajić, J. (2013). The influence of lecithin from different sources on crystallization and physical properties of non trans fat. *Journal of Texture Studies,* 44, 450-458.

[5] Szuhaj, B. F. (2005). Lecithins. In: F. Shahidi (Ed.), *Bailey's Industrial Oil and Fat Products* (pp. 371), New Jersey, John Wiley & Sons.

[6] Yang, M., Zheng, C., Zhou, Q., Huang, F., Liu C., & Wang H. (2013). Minor components and oxidative stability of cold-pressed oil from rapeseed cultivars in China. *Journal of Food Composition and Analysis,* 29, 1-9.

[7] Arnold, G., Schuldt, S., Schneider, Y., Frederichs, J., Babick, F., Werner, C., & Rohm, H. (2013). The impact of lecithin on rheology, sedimentation and particle interactions in oil-based dispersions. *Colloids and Surfaces A: Physicochemical and Engineering Aspects,* 418, 147-156.

[8] Fernandes, G. D., Alberici, R. M., Pereira, G. G., Cabral, E. C., Eberlin, M. N., & Barrera-Arellano, D. (2012). Direct characterization of commercial lecithins by easy ambient sonic-spray ionization mass spectrometry. *Food Chemistry,* 135, 1855–1860.

[9] Radujko, I., Jurić, J., Pajin, B., Omorjan, R., Šereš, Z., & Šoronja Simović D. (2011). The influence of combined emulsifier 2 in 1 on

physical and crystallization characteristics of edible fats. *European Food Research and Technology*, 232, 899-904.

[10] Torbica, A., Pajin, B., Omorjan, R., Lončarević, I., & Tomić J. (2014). Physical properties of chocolate with addition of Cocoa Butter Equivalent of moderate hardness. *Journal of the American Oil Chemists Society,* 91, 39-48.

[11] Pajin, B., Karlović, Đ., Omorjan, R., Sovilj, V., & Antić D. (2007). Influence of filling fat type on praline products with nougat filling. *European Journal of Lipid Science and Technology,* 109, 1203–1207.

[12] Ramadan, M.F. (2008). Quercetin increases antioxidant activity of soy lecithin in a triolein model system. *LWT-Food Science and Technology,* 41, 581-587.

[13] Cabezas, D.M., Diehl, B., & Tomas, M.C. (2009). Effect of processing parameters on sunflower phosphatidylcholine-enriched fractions extracted with aqueous ethanol. *European Journal of Lipid Science and Technology,* 111, 993-1002.

[14] Bueschelberger, H. G. (2004). Lecithins. In R. J. Whitehurst (Ed), *Emulsifiers in Food Technology* (pp.18). Northampton: Blackwell.

[15] Zarić, D., Pajin, B., Lončarević, I., Šereš, Z., Dokić, Lj., & Šoronja-Simović, D. (2012). The impact of manufacturing process on the content of hard triglycerides, hardness and thermal properties of milk chocolate. *Chemical Industry,* 66 (5) 735–741.

[16] Pajin, B., Dokić, Lj., Zarić, D., Šoronja-Simović, D., Lončarević, I., & Nikolić I. (2013). Crystallization and rheological properties of soya milk chocolate produced in a ball mill. *Journal of Food Engineering,* 114 (1), 70–74.

[17] Pajin, B. (2014). *Technology of chocolate and cocoa products.* Novi Sad: Faculty of Technology University of Novi Sad.

[18] Petković, M., Pajin, B., & Tomić J. (2013). Effects of temperature and mixer speed rotation on rheological properties of spreads with maltitol. *Journal of Food Process Engineering,* 36, 634–644.

[19] Joshi, A., Paratkar, S., & Thorata, B. N. (2006). Modification of lecithin by physical, chemical and enzymatic methods. *European Journal of Lipid Science and Technology,* 108, 363–373.

[20] Oke, M., Jacob, J.K. & Paliyath, G. (2010). Effect of soy lecithin in enhancing fruit juice/sauce quality. *Food Research International,* 43, 232–240.

[21] IOCCC (2000). *Viscosity of Cocoa and Chocolate Products - Analitycal Method* 46. Available from CAOBISCO, rue Defacqz 1, B-1000 Bruxelles.

In: Rapeseed
Editor: Monica White
ISBN: 978-1-63484-227-3

Chapter 4

LIPIDOMICS, PROTEOMICS AND PROFILING OF LOW MOLECULAR WEIGHT COMPONENTS IN *BRASSICA NAPUS* PLANT BY MASS SPECTROMETRIC BASED ANALYTICAL METHODS

Bojidarka Ivanova* and Michael Spiteller
Lehrstuhl für Analytische Chemie, Institut für Umweltforschung,
Fakultät für Chemie und Chemische Biologie, Universität Dortmund,
Dortmund, Nordrhein-Westfalen, Deutschland

ABSTRACT

Rape plant is economically important oilseed crop, contributing up to 15% of oil production worldwide. European Union Directive 2009/28/EC mandatory determines for member States 10% sharing of energy from renewable sources by 2020, thus concentrating much effort in implementation of rapeseed fatty acid methyl esters as biodiesel, yielding to established standards in blends with petrodiesel, having benefits over diesel fuel such as decreasing in CO_2 and smoke density (European Standards EN 590). But rapeseed has beneficial nutritional properties, providing most important nutrition for human daily life: energy, essential

* Corresponding author: Tel.:+ 49 231 755 4089, Fax. + 49 231 755 40 84; E-mail: B.Ivanova@infu.uni-dortmund.de; B.Ivanova@web.

fatty acids, phytosterols, and tocopherols. It has high protein content; over 200 flavonoids; and secondary metabolites showing antioxidant activity, prevention from cardiovascular diseases and cancer. The oil is 46% of dry weight of seed. Meal has 40% proteins, 35% fibers, 8.5% minerals and 1% lipids. Rape plant, thereby, is a fertilizing product of great importance for agriculture, too. We have chosen to emphasis on rapeseed as a food, calling attention to mass spectrometry for lipidomics, proteomics and low molecular weight analysis. The great applicability of mass spectrometry has been governed by high complexity of the foodstuff matrix, containing thousand components from biomacromolecules (12–21 kDa) to low molecular weight analytes in concentrations from macro-components (acylglycerols, > 94%) to traces (cytokinins, < $pmol.g^{-1}$). The analysis of agricultural pesticides is a challenging research task, too, due to lipophylic properties at trace levels, difficult their extraction. But, mass spectrometric methods have superior instrumental characteristics over a large scale of methods. These are: (*i*) instrumental flexibility and different ionization approaches, including off– and on-line hybrid operation; (*ii*) high–to–ultrahigh resolving power allowing analysis in multicomponent mixture. Particularly for rapeseed oil at resolution $m/\Delta m_{50\%} > 350\ 000$ and a mass accuracy < 1 ppm it has been achieved detail analysis of chemical content by electrospray ionization Fourier transform ion cyclotron resonance mass spectrometry without sample pretreatments; (*iii*) high accuracy and precision; (*iv*) low concentration limits of detection of *attomol* levels; (*v*) analytical information from macromolecules with masses higher than 50 kDa to low molecular weight analytes; and (*vi*) 3D imaging for assay from isolated living cell to whole organs and bodies. The chapter can be regarded as a comprehensive summary about contribution of mass spectrometric methods to analysis of rapeseed, reflecting their irreplaceability to the field of "Food control."

Keywords: rapeseed, lipidomics, proteomics, metabolomics, mass spectrometry, analytical chemistry

ACRONYMS

A	Arachnid acid
APCI	Atmospheric pressure chemical ionization
B	Behenic acid
CEC	Capillary electro-chromatography
C	Chemical ionization
DGs	Diacylglycerols

EI	Electron ionization
EDTA	Ethylenediamine tetraacetic acid
ESI	Electrospray ionization
G	Gadoleic acid
GC	Gas chromatography
GC–C–SIRMS	Gas chromatography–combustion-stable isotope ratio mass spectrometry
CE	Capillary electrophoresis
GPC	Gel permeation chromatography
DART	Direct analysis in real time (DART)
EI	Electron impact (mass spectrometry)
ESI FT–ICR MS	Electrospray ionization Fourier transform ion cyclotron resonance mass spectrometry
Fs	Fluorescence (detection)
FAB	Fast atoms bonbardment (mass spectrometry)
GLC–MS	Gas–liquid chromatography mass spectrometry
HPLC	High-performance liquid chromatography
HPSEC	High-performance size-exclusion chromatography
HS–SPME	Head-space–solid phase micro-extraction
ICP–MS	Inductively coupled plasma mass spectrometry
IR	Infrared
L	Linoleic acid
Li	Lignoceric acid
LLE	Liquid–liquid extraction
LMW	Low molecular weight (analytes)
Ln	Linolenic acid
MAE	Microwave-assisted extraction
MGs	Monoacylglycerols
MPSE	Magnetic solid-phase extraction
Mo	Heptadecanoic acid
MSPD	Matrix solid-phase dispersion
NARP	Non–aqueous reversed–phase
NMR	Nuclear magnetic resonance
NARP–HPLC	Non-aqueous reversed-phase high-performance liquid chromatography
O	Oleic acid
P	Palmic acid
PA	Phosphatidic acid
PC	Phosphatidylcholine

PE	Phosphatidylethanolamine
PG	Phosphatidylglycerol
PI	Phosphatidylinositol
PMME	Polymer monolith microextraction
Po	Palmitoleic acid
PS	Phosphatidylserine
QuEChERS	Quick, easy, cheap, effective, rugged, and safe
RI	Relative intensity
S	Stearic acid
SFC	Supercritical fluid chromatography
SPE	Solid–phase extraction
TLC	Thin layered chromatography
TGs	Triacylglycerols
TOF	Time-of-flight (analyzer)
UPLC	Ultra–high performanceliquid chromatography
UV	ultraviolet (detection).

INTRODUCTION

We should note here, in the "introduction" part of this chapter, like in "summary" one about the wide applicability of oilseed rape fatty acids methyl esters as reniewvable source of energy, because of enormous research efforts have been devoted to development of "Clean technologies" blending the petrodiesel, on the base of oilseed crops, thus optimizing the naturally occurring sources at minimum environmental impact [1–9]. Particularly rapeseed based biodiesel has great potential to replace the fossil fuels, because of there have already been discovered crucial genetic, biotechnological and process chemical engineering technologies improving the quality and the applicability of the rapeseed oil for such purposes [1–9]. There is evidenced that the use of rapeseed based biodiesel is characterized with decreasing in CO_2 and smoke density, thus rescuing the emission of greenhouse gases. Particularly within the frame of the member States of European Union have been concentrated innovation programs and investment in standard rapeseed biodiesel (European Standards EN 590) according to European Union Directive 2009/28/EC, determining intensive development and implementation of new "Clean technologies." Furthermore the 1st generation "Clean" fuels, based on biodiesel is currently commercially available [1–9]. Despite the fact that the technological cycle leading to the implementation of rapeseed oil products as biodiesel has broadly involved analytical steps using mass spectrometric based protocols for chemical analysis [10–26] as among the robustest instrumentation currently available, in this chapter, we will introduce the great capability of the methods of mass spectrometry to "Food control," describing their contribution to qualitative, quantitative and structural analysis of the chemical composition of *Brassica napus L.* and its products, including the food technological steps for rapeseed edible oil production. From this perspective we shall argue that some detail attention on the available research contributions clearly highlighted the beneficial nutritional properties of the rapeseed products both to humans and animals, determining rapeseed oil as among the healthy oil product for the humans. Particularly important is to highlight it antioxidant, antimicrobial and anti-fungal activity, in addition to prevention from cardiovascular diseases and cancer [27, 28]. Surprisingly, therefore, in the case of *Brassica napus L.* plant there has a significant competition between developments in high healthy for the humans food– and, in parallel, "Clean" technologies. It should, of course, be mentioned here that our chapter, having essentially character of "chapter-review," addresses contributions of the methods of mass spectrometry to significantly important technologies at an industrial scale, reflecting the attention to known reports, but in parallel, emphasizing on great prospectives for further both laboratory and industrial scale analytical elaborations, implementing the mass spectrometric based methods to many interdisciplinary branches involving chemistry, chemical and process engineering, biotechnology, genetics and the fourth. For this reason we have chosen rather an "object-oriented" structure of the content of the chapter, than to summarize contributions to each of the different methods of the mass spectrometry applied towards rapeseed–based analytical objects.

1. Rapeseed Lipidomics

The rapeseed lipidomics has essentially been associated with macro-component analysis of tri–, di– and monoacylglycerols, where TGs content can vary $\in$ 94–98% in crude/cold pressed oil to 99% in the corresponding refined product [29]. Despite the fact that the fatty acids belong to the LMW analytes, accounting to their molecular masses, we have discussed their analysis separately because they are the main components of the rapeseed food product. So, series papers report efforts in elaboration of MS–based protocols for lipidomics in rapeseed [30–53]. Particularly challenging from an analytical point of view is quantitation of *cis/trans*–isomers and positional isomers of unsaturated fatty acids [49]. Furthermore, fatty acid content can vary within the frame of different maturation stages [36]. The structural similarity of major fatty acid components in rapeseed oil, thereby, requests sensible and reliable analytical method for structural analysis and quantitation. GC could be used for analysis of MGs, however, main drawbacks, connected with determination of non–volatile compounds have occurred. The TGs screening has involved short capillary column operating at high temperature (T = 350°C). The employment of TLC as a preliminary stage of sample preparation has shown low selectivity and resolution accompanied with numerous and costly operation stages [49, 54, 55]. Normal phase HPLC coupled with MS has been successfully used for quantitation of TGs, DGs and MGs, respectively [11, 55–69], nevertheless that selectivity of *cis/trans*– and different positional isomers is relatively low. Among developed HPLC approaches for preliminary separation of TGs there are widespread used NARP–HPLC and silver mode HPLC [31, 49, 58–60]. The last method has been employed in separation of positional isomers of TGs [11]. The GLC approach needs a preliminary chemical derivation step converting fatty acids into their methyl ester derivatives; while a carbon isotope labeling requires whole sample to be preliminary combusted [70]. So that, despite, numerous reports and successful analyses by chromatographic separation steps, all these approaches are significantly time and labor consuming in addition to their relatively high cost. Furthermore none of them have high selectivity to individual acylglycerols [70]. On the other hand, TGs do not absorb irradiation energy within 200–400 nm, and there is unable to achieve quantitation by Fs or UV–detection. CEC, representing a hybrid methods between the capillary electrophoresis and micro–LC, approach has reserved much attention as a fast separation technique operating at a higher resolution, than LC–one as well as providing more complete analyte profiling than the LC–approach, applicable to lipidomics [62]. By contrast, MS methods provide qualitative, quantitative and structural information about individual analytes in complex mixtures like rapeseed oil. Early studies on the MS fragmentation of TGs in oils [67] have employed the CI– and EI–approaches. The first one has yielded to abundance $[M–H]^-$, operating at negative mode, ion without to produce interference $[M–35]^-$ ion. The positive operation mode has resulted to $[M+H]^+$– and $[M+NH_4]^+$– adducts, in addition to $[M+H–RCO_2H]^+$ ion. The EI–ionization has revealed formation of $M^{\bullet+}$ cation–radical in addition to corresponding fragmentation patterns operating under both positive or negative modes. Among powerful MS–methods, widely used are APCI– and ESI–MS [11].

Nevertheless that a low abundance of $[M+H]^+$ ion for different TGs operating at a positive mode by APCI–MS, have been observed, $[M+H–RCOOH]^+$ ion usually exhibits a relatively high intensity enables quantitation of the corresponding analytes. More complex MS picture has yielded ESI–MS based method, showing series $[M+Na]^+$, $[M+K]^+$ and $[M+NH_4]^+$ adducts of analyte TGs. Comparing with APCI–$(MS)^n$ analysis, ESI–$(MS)^n$ one also shows complex fragmentation patterns due to stabilization of last mentioned ions (Table 1). In spite of those drawbacks to chemical derivation, and in parallel to methylation of fatty acid derivatives as an approach to their identification it has also been utilized approach of formation of sulfide adducts as a tool for determination of double bond position in molecular scaffolds of unsaturated acids, employing TLC and GC–MS [30]. The procedure involves treatment with dimethyl disulfide yielding to methyl 9,10-bis(methylthio)octadec-12-enoate (Isomer I) and 12,13-bis(methylthio)octadec-9-enoate (Isomer II) treating acid L (18:2, $D^{9,12}$), respectively (Scheme 1). The MS patterns obtained allow unambiguous distinguishing of both isomers showing peaks at *m/z* 217.1240/171.1199 (Isomer I) and 257.1554/131.0897 (Isomer II), respectively.

Table 1. Relative abundance of $[M+H]^+$, $[M+Na]^+$, $[M+K]^+$ and $[M+NH_4]^+$ ions of main TGs obtained by ESI– and APCI–MS methods [11]

	APCI		
TGs	$[M+H]^+$		
PPP	0.03		
PoPoPo	6.3		
OOO	7.1		
LLL	100		
LnLnLn	100		
	ESI		
TGs	$[M+Na]^+$	$[M+K]^+$	$[M+NH_4]^+$
PPP	100	71.7	8.1
PoPoPo	-	-	
OOO	26.9	100	13
LLL	18.1	100	3.6
LnLnLn	13.1	100	3.1

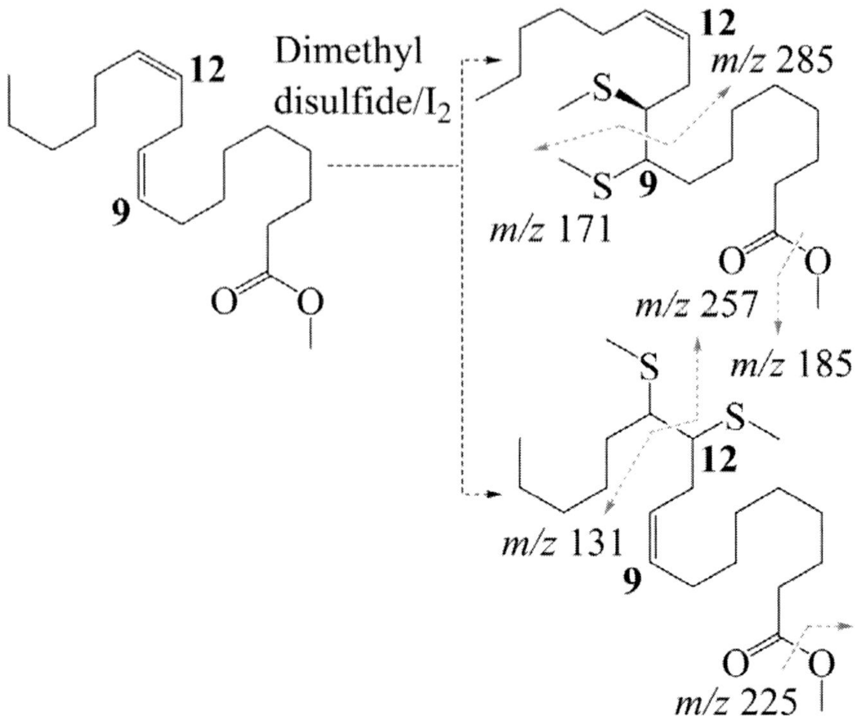

Scheme 1. Chemical derivatization of L (18:2, $\Delta^{9,12}$) by dimethyl disulfide; MS data [30].

The implementation of MALDI–MS for analysis of TGs is still scarce, despite reported prominent result [11]. The main reason to a rather lack of MALDI–MS based studies for analysis of major component in oils can be associated with a wide trend to implementation of this method for biological macromolecular screening of living cell, organs and bodies determining its application rather as method for clinical diagnostics and medical research [71–74], than to areas such as, for example, food chemistry and "Food control" [75]. A significant progress, which has been achieved is associated with ESI FT–ICR MS study of rapeseed oil, showing that operating at ultrahigh MS resolving power of $m/\Delta m_{50\%} > 350\ 000$ and mass accuracy <1 ppm it has been identified in details chemical components without any incorporation of extraction or separation stages as well as any chemical derivation procedure [77]. These crucial results shed light into prospectives for development of MS based methods for direct assay of oil foodstuffs allowing, obtaining of meaningful analytical information at high accuracy and precision reducing the systematic and random errors of measurements typical for multi-stage sample pretreatment approaches, in addition to an overcoming drawbacks to many methods, using preliminary sample pretreatments. But, particularly prominent among MS–based methods for such as purposes is namely the MALDI–MS, involving Orbitrap analyzer, which like FT–ICR method has ultra–high resolving power and superior instrumental characteristics mentioned still in the "summary" part to this chapter. Moreover it allows direct assay without any preliminary sample manipulations, involving concentration, extraction, purification and the fourth [75]. More recently has been discussed the instrumental advantages of MALDI–MS for environmental analysis of complex multicomponent mixtures such as aquatic and terrestrial soil and sediments, making a brief parallel with it applicability to foodstuffs analysis, thus aiming to highlight that, in spite of, already mentioned outstanding contributions of MALDI–MS to clinical diagnostics and medicinal practice, in fact, the great ability the method encompasses a large range of molecular masses from LMW analytes to macromolecules having weights up to 100 kDa [71–74].

For purposes of a structural elucidation of unknown oils, there has been elucidated fragmentation patterns of common TGs, DGs and MGs (Scheme 2) as a base to identify major components in rapeseed oil [11, 33, 42, 77].

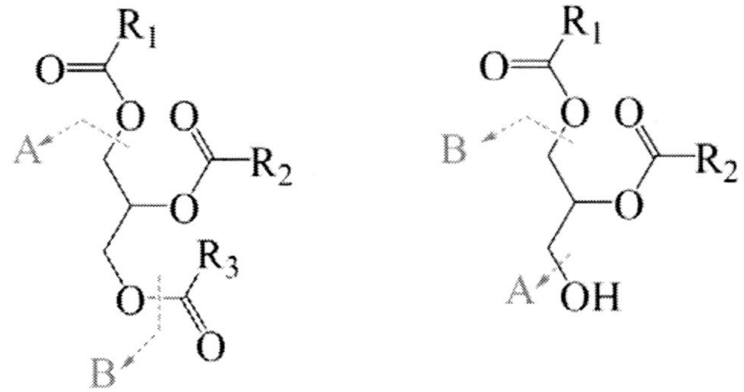

$A = [M+H-R_iCOOH]^+$ $\quad$ $A = [M+H-H_2O]^+$

$B = [M+H-R_iCOOR_iCO]^+$ $\quad$ $B = [M+H-R_iCOOR_iCO]^+$

Scheme 2. (Continued).

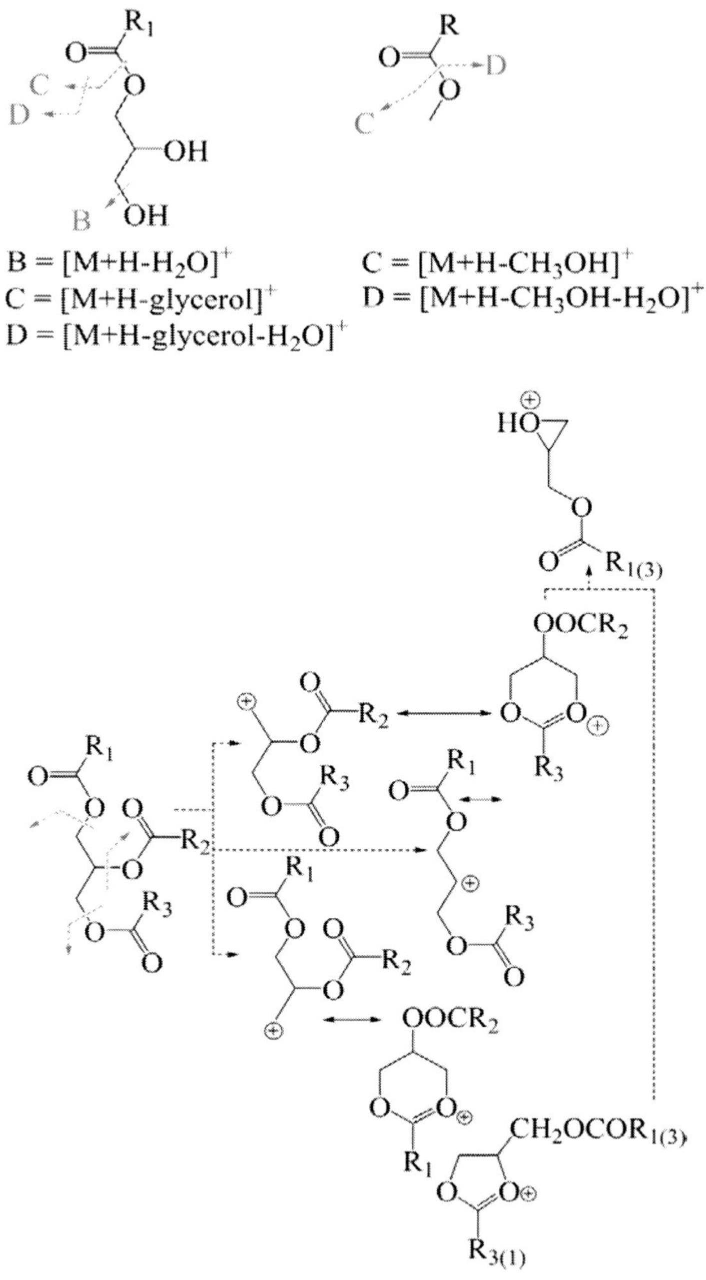

Scheme 2. Fragmentation scheme of major acylglycerols and their methyl esters [11, 33, 42].

Disusing analytical advantages of the method of the mass spectrometry not only for quantitation, but also for structural analysis, it is important to pointed out their advantages in distinguishing *cis/trans*–isomers of unsaturated fatty acids. From an analytical point of view as it has been mentioned above this research task appears challenging, moreover, the widely employed chromatographic methods have exhibited little selectivity towards those isomers. By contrast a FAB–MS study has shown that both isomer types have shown different intensity ratios of MS isotope profile [69]. Thus, relative intensity ratio of MS peaks at *m/z* 181/182, 209/210 and 237/238 in *cis/trans*–9–octadecenoic, 11–octadecenoic and 13–docosenoic acids are: RI^{181}/RI^{182} = 1.19 (*cis*–9–octadecenoic acid), RI^{181}/RI^{182} = 0.67 (*trans*–9–octadecenoic acid), RI^{209}/RI^{210} = 1.31 (*cis*–11–octadecenoic acid), RI^{209}/RI^{210} = 0.89 (*trans*–11–octadecenoic acid), RI^{237}/RI^{238} = 1.28 (*cis*–13–docosenoic acid) and RI^{237}/RI^{238} = 0.85 (*trans*–13–docosenoic acid), respectively.

The employment of soft–ionization MS methods [71–76] briefly discussed in previous paragraph has resulted to detail component identification of rapeseed oil, starting with major fatty acids [*w/w*%]: O (C18:1), *m/z* 339 (62.1%); L (C18:2) *m/z* 337 (19.8%); Ln (C18:3) *m/z* 335 (8.5%); P (C16:0) *m/z* 313 (4.6%); S (C18:0) m/z 341 (1.7%); (C20:0) (0.6%); G (C20:1) 367 (1.6%); (C22:0) (0.3%) and (C22:1) (0.8%) [33]. Total number of TGs 38 has been identified [31], using chromatographic peak ratios, MS peak values of molecular ions [M+H]+ and retention times (τ_R): LnLnLn (Area, < 0.05; *m/z* 873; τ_R = 81.5 mins); LLnLn (Area, < 0.05; *m/z* 875; τ_R = 87.9); LLLn (Area, 2.7; *m/z* 877; τ_R = 94.0); OLnLn (Area, 3.7; *m/z* 877; τ_R = 94.9); LnLnP (Area, 2.3; *m/z* 851; τ_R = 96.7); LLL (Area, 1.1; *m/z* 879; τ_R = 99.7); OLLn (Area, 5.7; *m/z* 879; τ_R = 100.8); LnLP (Area, 0.7; *m/z* 853; τ_R = 102.6;); OLL (Area, 8.0; *m/z* 881; τ_R = 105.8); OLnO (Area, 11.1; *m/z* 881; τ_R = 106.9); LLP (Area, 1.1; *m/z* 855; τ_R = 107.5); OLnP (Area, 2.3; *m/z* 855; τ_R = 108.4); PLnP (Area, 0.06; *m/z* 829; τ_R = 110.1); LOP (Area, 5.4; *m/z* 857; τ_R = 114.2); PLP (Area, 0.7;831; *m/z* 116.2; OOMo (Area, 0.2; *m/z* 871; τ_R = 116.3); GLO (Area, 1.2; *m/z* 911; τ_R = 118.4); OOO (Area, 16.8; *m/z* 855; τ_R = 119.0); SLO (Area, 2.1; *m/z* 855; τ_R = 120.6); OOP (Area, 4.3; *m/z* 859; τ_R = 120.9); POP (Area, 0.8; *m/z* 833; τ_R = 122.6); GLS (Area, 0.5; *m/z* 913; τ_R = 123.9); GOO (Area, 1.3; *m/z* 913; τ_R = 124.6); ALO (Area, 0.3; *m/z* 913; τ_R = 126.1); GOP (Area, 0.8; *m/z* 887; τ_R = 126.3); SOO (Area, 1.2; *m/z* 887; τ_R = 126.7); SOP (Area, 0.5; *m/z* 861; τ_R = 128.7); GOG (Area, 0.3; *m/z* 941; τ_R = 129.7); LiLL (Area, 0.3; *m/z* 967; τ_R = 130.8); BOL (Area, 0.5; *m/z* 941; τ_R = 131.4); AOO (Area, 0.4; *m/z* 915; τ_R = 132.0); SOS (Area, 0.07; *m/z* 889;

τ_R = 133.8); LiOL (Area, 0.06; *m/z* 969; τ_R = 136.0); BOO (Area, 0.1; *m/z* 943; τ_R = 136.9); GGS (Area, < 0.05; *m/z* 943; τ_R = 139.0) and LiOO (Area, < 0.05; *m/z* 971; τ_R = 141.5), respectively [31, 42].

The MS spectra of saturated TGs such as for example tritripalmitin operating at tandem MS^3 mode has shown $[M+NH_4]^+ \rightarrow [(M+NH_4)-NH_3-H_2O]^+$ fragmentation, while the DGs such as 1,3-dihexadecanoylglycerol exhibits abundance $[(M+NH_4)-NH_3-H_2O]^+$ ion with MS peak at *m/z* 552 (100%) [42].

A step-wise mechanism of fragmentation has been proposed studying the MS2 pattern of 1,3-dihexadecanoylglycerol, showing high abundance [(M+NH4)–NH3]+ ion. It has MS peak at *m/z* 569 [42]. The MS^2 analysis of MGs also indicated formation of abundance $[M+NH_4]^+$ adducts having relative intensities of 100%, in addition to a low–abundance MS peaks of $[(M+NH_4)-NH_3]^+$ and $[(M+NH_4)-NH_3-H_2O]^+$ ions.

As an important analytical task can be regarded the determination of the cyclic fatty acids, obtained as a result of intra-molecular formation of cyclic unsaturated fatty acids in rapeseed oil [43]. Because of to some of these cyclic products have been reported adverse physiological activity there are focused much effort in development of quantitative protocols for their analysis during the refinement and frying processes, where these derivatives ha mainly occurred in the oil product. Thus 11,12-methylene-octadecanoic acid has been identified, as a cyclic product performing a comprehensive correlative GC–MS analysis of their derivatives 11,12-methylene-octadecanoate methyl ester and .ass spectrum of the picolinyl derivative of 11-nonadecenoic acid [43]. It has been shown however, that this fatty acid derivative exists also in crude oil extract and therefore, it cannot be assigned to oil-processing derivatives.

The successful identification of TGs, generally, applicable to rapeseed oil products, through their Li – adducts in the gas – phase has been demonstrated [68]. The ESI–MS method has been involved in the experimental design, showing formation of abundance $[M+Li]^+$ ions. These ions has yielded abundance fragment ions of type $[M+Li-(R_nCO_2H)]^+$ and $[M+Li-(R_nCO_2Li)]^+$, respectively. Each fatty acid substituents can be easy identified using the abundance fragment ions $[R_nCO_2H+Li]^+$, $[R_nCO]^+$, and $[[R_nCO]^+-18]$, respectively.

In this sub–section of the chapter, also, we have discussed adulteration of rapeseed oil with cheaper oils, representing a significant challenge from an analytical point of view, due to overlapping of fatty acid compositions in many different oil crops [77].

Despite this challenge GC–SIRMS and DART–HRMS approach has yielded to successful profiling of different oils determining fatty acid MS fingerprinting profiles. Usually the analysis has involved carbon isotope labeling [75].

2. Rapeseed Proteomics

The food plant protein has been regarded as an important source of peptides and amino acids with various biological functions. Particularly rapeseed proteins have received great attention due to proposed anti-tumor activity, in addition, to a broad spectrum of anti-fungal one [78–82]. The main protein content in meal is 40%, while totally 20–25% proteins have occurred in dry seed weight [81, 82]. In this respect, it has been focused significant effort in obtaining of rapeseed protein adequate for human consumption [78]. The main difficulties are associated with presence of low molecular weight analytes such as glucosinolates and erucic acid (Section 3.6). The studies in this direction have shown that in rapeseed have mainly two families of storage proteins, i.e., cruciferins and napins [83]. The first group belongs to 12S–glubolin class and represents ∈ 15–45% of total content. The second family proteins belongs to 2S albumins, having molecular weights ∈ 12500–14500 Da [79, 84–92]. After maturation there have been identified two peptide chains (chains A and B, Figure 1) having masses 3.8–4.5 and 8.4–10 kDa joined with S–S disulfide bridges. Due to numerous napin genes and various proteolytic cleavage it has been found existing of a large set of different isoforms.

The employment of MALDI–TOF and FAB mass spectrometry has resulted to identification of a large number of those isoforms summarized in Table 2 [93–95].

The 3D structure of one of those forms called BnIb has been determined using NMR (Figure 2) [79]. These are still scarce findings on the 3D molecular structures of rapeseed proteins however are the base for further determination of the whole range of biological function of the napin type 2S albumins, which generally have antimicrobial activity, in addition, to already mentioned anti-fungal one [79, 93].

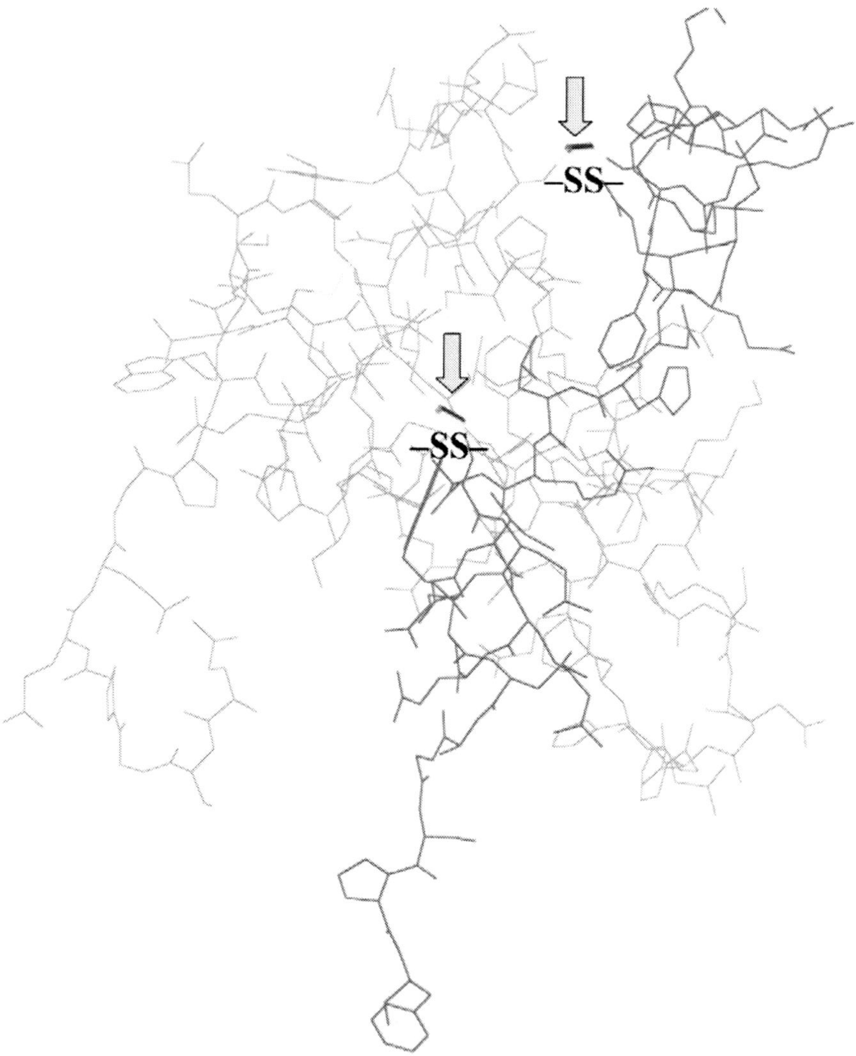

Figure 1. Molecular structure and S–S linkage of napin protein isoform BnIb according [79].

Table 2. Proteomics by MALDI–TOF and ESI–MS/MS of short and long napin chains

Sequence/protein identity	Position/gene ID	M/z	Refs.
Pro-Ala-Gly-Pro-Phe	38–42	488.2	[28]
IP...QSGGGPS	7-39	3991.4	[93]
RIP...QSGGGPS	6-39	4147.6	[93]
GPFRIP...QSGGGPS	3-39	4448.5	[93]
AGPFRIP...QSGGGPS	2-39	4520.2	[93]
SAGPFRIP...QSGGGPS	1-39	4604.3	[93]
IP...QSGSGPS	7-39	3991.4	[93]
RIP...QSGSGPS	6-39	4147.6	[93]
GPFRIP...QSGSGPS	3-39	4448.5	[93]
AGPFRIP...QSGSGPS	2-39	4520.2	[93]
PAGPFRIP...QSGSGPS	1-39	4617.4	[93]
QAGPFRIP...QSGSGPS	1-39	4631.4	[93]
IP...QSGSGPS	7-39	4019.1	[93]
RIP...QSGSGPS	6-39	4175.8	[93]
GPFRIP...QSGSGPS	3-39	4477.6	[93]
PAGPFRIP...QSGSGPS	1-39	4645.3	[93]
RIP...GGGSGPS	6-41	4299.3	[93]
PAGPFRIP...GGGSGPS	1-41	4769.1	[93]
< QAGPFRIP...GGGSGPS	1-41	4783.6	[93]
< Q...KTMPGP	6'-89'	9567.2	[93]
PQGPQQ...KTMPG	1-88'	9995.3	[93]
LSAGPNGGSIAK		16794	[97]
LTFVEDGETK			[97]
Luminal binding protein	30693966	164	[99]
Luminal binding protein	30693962	243	[99]
Putative heat shock protein partial	37718900	131	[99]
Dipeptidyl peptidase IV	30794015	225	[99]
Putative inorganic pyrophosphatase	26451240	137	[99]
Luminal binding protein	19805	82	[99]
PQGPQQ...KTMPGP	1 '-89'	10093.8	[93]
PQGPQQ...KTMPGPS	1 '-90'	10179.8	[93]
PQGPQQ...KTMPG	1 '-88'	10265.1	[93]
PQGPQQ...KTMPGP	1-89'	10358.7	[93]
PQGPQQ...KTMPGPS	1 '-90'	10447.1	[93]
< Q... KTMPGPG	6'-90'	10242.8	[93]
GPQQ.. .KTMPGPG	3'-90'	10542.4	[93]
< QGPQQ.. .KTMPGPG	2'-90'	10644.7	[93]
< QQGPQQ... KTMPGPG	1 '-90'	10778.4	[93]
< QQQGPQQ...KTMPGPG	0'-90'	10908.5	[93]

Table 2. (Continued)

Sequence/protein identity	Position/gene ID	M/z	Refs.
< Q...KTMPG	6'-88'	9710.2	[93]
< Q...KTMPGP	6'-89'	9806.8	[93]
< Q...KTMPGPS	6'-90'	9893.2	[93]
PQGPQQ...KTMPG	1'-88'	10231.9	[93]
PQGPQQ...KTMPGP	1'-89'	10331.3	[93]
PQGPQQ...KTMPGPS	1 '-90'	10417.5	[93]
PQSPQQ...KTMPG	1'-88'	10264.1	[93]
PQSPQQ...KTMPGP	1'-89'	10361.2	[93]
PQSPQQ...KTPMGPS	1 '-90'	10447.2	[93]
GDAAPSEEQLK			[97]
ALVTDADNLTPK			
Phosphoglycerate mutase	23297457	413	[99]
Cytoplasmic aconitate hydratase	4586021	439	[99]
Phosphoglycerate mutase	18391066	444	[99]
Phosphoglycerate mutase	18391066	645	[99]
Putative elongation factor	29824421	162	[99]
Phytocystatin	33309455	97	[99]

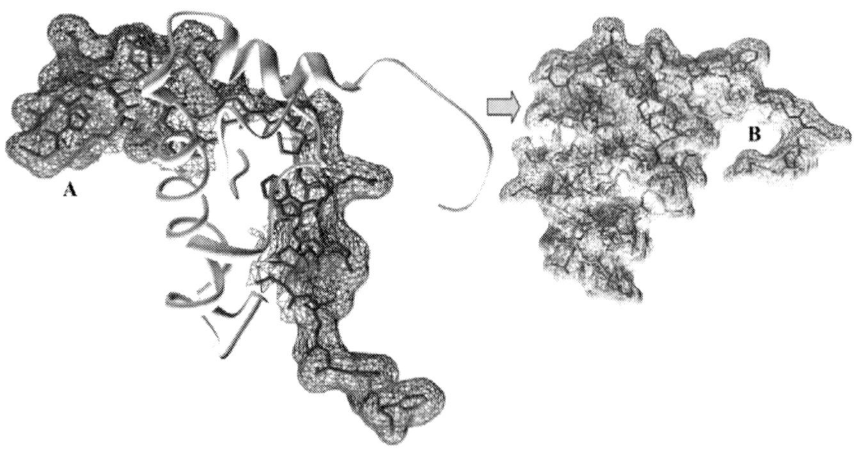

Figure 2. 3D structure of napin protein isoform Bnlb according [79].

Moreover these proteins are susceptible to a genetic manipulations, thus appearing excellent templates for the genetic engineering in improvement of their biological role and function as well as the healthy properties of the grain crop. Cyclic Zn–containing metal–organics of small peptides such as Ser–Met and Asn–Cys–Ser in rapeseed oil has determined in rapeseed employing ESI–MS [96]. Thus the summary of proteomic studies of *Brassica napus* plant has been shown in Table 2, encompassing the contributions [97–142], in addition to those ones already discussed above.

3. Profiling of Low Molecular Weight Components in *Brassica Napus* Plant and Its Products

It was pointed out above that the analysis of fatty acids from an analytical chemical point of view, employing MS based methods essentially represents analysis of LMW–analytes. But because of the fatty acids are the major components in the rapeseed crop, the lipodimics has been discussed as a separate section 1. The analysis of LMW analytes in rape plant flowers, hulls, seed and roots as well as the rapeseed oil [143–210] will be discussed in the next sub-sections in this section 3.

3.1. Analysis of *Brassica Napus* Flowers and Hulls

The MS analysis of *Brassica napus* hulls employing the GC–, LC–MS and FAB–MS methods has shown that they have significant amount of carbohydrates. MS analysis of extracts by aqueous ammonium oxalate medium has shown the major content of D–galacturonic acid (76%), D–galactose (2–3%), L–arabinose (8–9%), D–xylose (2%) and L–fucose (1%) [151]. While, partial acetolysis of xyloglucan, followed by a O–deacetylation has resulted to series monosacchadides: 2,3,4-$(CH_3)_3$–xylose (*m/z* 117 (118), 161 (162)); 2,3,4-$(CH_3)_3$–fucose (*m/z* 117 (118), 131, 161 (162), 175); 3,4-$(CH_3)_2$–xylose (*m/z* 117, 189(190)) and 3,4,6-$(CH_3)_3$–galactose (*m/z* 161, 189 (190)) [152] In addition oligosaccharides and series phenolic acids have been also determined in flowers and hulls of rape plant [143, 153]. The GLC–MS analysis of hulls has shown the presence of: p–hydroxy benzoic acid (0.2 mg/100 g flower); vanillic acid (1.2); protocatechuic acid (16.4), *p*–coumaric and *cis/trans*–ferulic acids (trace concentration levels); caffeic acid (4.3) and trans-sinapic acid (2.4), thus determining a total chemical content of those derivatives of 24.5 mg/100 g flower [143, 202]. The malic, synapic and chlorogenic acids have been determined by mass spectrometry using the following sets of MS peaks and corresponding abundances of the molecular ions to each of the acids as following: *m/z* 423 (0.7), 335 (7.4), 245 (14.8), 233 (31.7), 189 (13.6), 175 (8.9), 147 (71.2) and 133 (11.9); 368 (93.6), 353 (38.1), 338 (87.0), 323 (17.0) and 279 (21.4); as well as 768 (2.8), 372 (3.0), 345 (47.3) and 307 (27.5), respectively [143].

3.2. Analysis of *Brassica Napus* Seed

The pyrolysis of rapeseed meal has yielded the proximate and ultimate components showing, moisture (10.58 wt%), volatile matter (67.31 wt%), fixed carbon (15.80 wt%), C (45.98 dry wt%), H (6.21 dry wt%), N (6.90 dry wt%), S (0.88 dry wt%), O (40.09 dry wt%), extractive (19.40 dry wt%), hemicellulose (41.40 dry wt%), lignin (4.99 dry wt%) and cellulose (28.59 dry wt%), respectively [211]. The protein content is 41.4%, fats 3.2%, ashes 7.6%, polyphenols 1.3%, phytic acid 1.2%, fibres and sugars 45.3%, respectively [212]. The significant protein content having healthy for humans biological activity (section 2). In parallel, however, the seed has also series antinutritives, where the fibres being among most abundance one [169]. The UPLC–MS/MS study [169, 170] of rapeseed fiber has identified series LMW analytes of phenolic and aldehyde chemical classes as summarized in Table 3. The analysis of vitamin K has been reported to [171]. In seeds there are determined employing HPLC–ESI–$(MS)^n$ numerous flavonoids, their glucosides, hydroxycynnamoyls, (–)–epicatechin, malic acid, procyanidin B2, and 1–synapoylgentiobiose, as well as non-glucosinolates (Scheme 3, Table 3) [172–174].

Table 3. LMW analytes MS/MS analysis

Analyte (Species)	t_R [min]	Fragmentation MS/MS^3/MS^4 data (Relative intensity,%)	[Ref.]
Protocatechuic acid	2.0	153 ($[M-H]^-$), 91 (5), 108 (31), 109 (100), 153 (44)	[169, 170]
4-Hydroxybenzoic acid	3.3	137 ($[M-H]^-$), 93 (100), 137 (18)	[169,170]
Salicylic acid	3.3	137($[M-H]^-$), 65 (8), 93 (100), 137 (54)	[169]
4-Hydroxybenzaldehyde	4.6	121($[M-H]^-$), 91 (5), 92 (56), 93 (12), 121 (100)	[169]

Analyte (Species)	tR [min]	Fragmentation MS/MS3/MS4 data (Relative intensity,%)	[Ref.]
5-Hydroxyferulic acid	5.2	209 ([M-H]$^-$), 121 (3), 149 (20), 150 (100), 165 (11), 194 (43), 209 (62)	[169]
p-Coumaric acid	5.7	163 ([M-H]$^-$), 91 (6), 93 (24), 117 (10), 119 (100), 163 (37)	[169]
Coniferyl alcohol	6.2	179 ([M-H]$^-$), 89 (3), 91 (5), 117 (5), 119 (3), 135 (13), 145 (23), 146 (100), 147 (12), 161 (15), 163 (14), 164 (61), 179 (42)	[169]
Ferulic acid	6.4	193.1 ([M-H]$^-$), 89 (20), 117 (6), 133 (18), 134 (100), 149 (10), 178 (18), 193 (49)	[169]
Sinapic acid	6.7	223.1 ([M-H]$^-$), 65 (3), 89 (4), 93 (30), 104 (3), 117 (3), 121 (59), 132 (3), 135 (11), 148 (3), 149 (100), 163 (12), 164 (46), 165 (6), 179 (7), 193 (37), 208 (24), 223 (53)	[169]
Sinapaldehyde	7.5	207.1 ([M-H]$^-$), 77 (5), 93 (15), 105 (9), 121 (17), 123 (3), 149 (29), 163 (3), 177 (100), 191 (5), 192 (75), 207 (50)	[169]
qn-3-O-sophoroside-7-O-glucoside	16.57	787.2 ([M-H]$^-$), 445 (23), 301 (85), 300 (100)	[172]
km-3-O-caffeoylsophoroside-7-Oglucoside	21.9	933.2 ([M-H]$^-$), 609 (100), 447 (11), 429 (29), 285 (5)	[172]
(−)-epicatechin	24.46	289.1 ([M-H]$^-$), 203 (100), 187 (39)	[172, 174]
(Putative) hydroxycinnamic acid derivative	49.54	949.3 ([M-H]$^-$), 419 (100), 223 (11)	[172]
1,6-Disinapoylglucoside	55.15	591.2 ([M-H]$^-$), 223 (45), 205 (100), 190 (41)	[172]
Quercetin-3-*O*-glucoside	12.3	463 ([M-H]$^-$), 301	[174]
Putative epicatechin derivative	13.3	386 ([M-H]$^-$), 306, 275, 259, 241, 208	[174]
Kaempferol-3-*O*-glucoside	19.0	447 ([M-H]$^-$), 609, 447, 285	[174]
Quercetin 3-*O*-diglucoside-7-*O*-glucoside	22.0	787 ([M-H]$^-$), 831, 625, 301	[174]

Table 3. (Continued)

Analyte (Species)	t_R [min]	Fragmentation MS/MS^3/MS^4 data (Relative intensity,%)	[Ref.]
Putative quercetin derivative	22.6	423 ([M-H]$^-$), 342, 301	[174]
Kaempferol-3-*O*-diglucoside-7-*O*-glucoside	23.6	771 ([M-H]$^-$), 609, 447, 285	[174]
Isorhametin-dihexoside-*O*-glucoside	24.8	801 ([M-H]$^-$), 639, 315	[174]
3-Methylsulphinylpropyl (glucoiberin)	2.9	358, 259	[89]
2-Propenyl (sinigrin)	3.8	278, 275, 259, 227, 196, 180, 162	[89]
3-Butenyl (gluconapin)	7/5	292, 275, 259, 227, 195, 194, 176	[89]
3-Methylthiopropyl	12.9	326, 275, 259, 228, 195	[89]
3-Indolylmethyl	21.4	275, 259, 251, 205	[89]
1-Methoxy-3-indolylmethyl	34.6	447, 466, 284, 259	[89]
Aflatoxin B1 (*Aspergillus flavus* and *parasiticus*)	15.9	313	[208]
Fumonisin B1 (*Fusarium verticillioides*)	18.8	722	[208]
Ochratoxin A (*Aspergillus ochraceus*)	23.6	404	[208]
Zearalenone (*Fusarium culmorum*)	23.6	317 ((-)-MS-mode)	[208]
Cholesterol	0.61	255, 129, 107	[144, 224]
Brassicasterol	0.69		[144]
Campesterol	0.81	255, 213, 129, 83, 73	[144, 224]
Δ^5-Avenasterol	1.12	412 M$^+$ (6), 386, 343, 296, 257, 129, 83	[144, 224]
Δ^7-Avenasterol	1.32	386, 343, 296, 281, 253	[144, 224]
2-(4-Hydroxyphenyl) ethanol	4.2	137 ([M-H]$^-$), 93 (8), 106 (4), 107 (7), 119 (24), 137 (100)	[169, 170]
Vanillic acid	4.3	167 ([M-H]$^-$), 65 (15), 91 (28), 108 (100), 123 (34), 152 (45), 167 (59)	[169]
Caffeic acid	4.5	179 ([M-H]$^-$), 89 (6), 107 (4), 134 (22), 135 (100), 179 (43)	[169]

Analyte (Species)	tR [min]	Fragmentation MS/MS3/MS4 data (Relative intensity,%)	[Ref.]
Syringic acid	5.0	197 ([M-H]⁻), 63 (3), 78 (17), 89 (50), 91 (50), 95 (38), 106 (38), 121 (53), 123 (47), 138 (54), 153 (50), 167 (17), 182 (54), 197 (100)	[169]
Vanillin	5.6	151 ([M-H]⁻), 92 (31), 108 (28), 136 (100), 151 (39)	[169]
Benzoic acid	5.9	121 ([M-H]⁻), 77 (19), 121 (100)	[169]
Syringaldehyde	6.3	181.1 ([M-H]⁻), 67 (11), 95 (19), 123 (50), 137 (4), 151 (100), 165 (8), 166 (79), 181 (42)	[169]
Sinapyl alcohol	6.6	209.1 ([M-H]⁻), 77 (5), 93 (11), 95 (3), 105 (6), 121 (8), 133 (18), 151 (46), 161 (59), 176 (18), 179 (100), 191 (7), 194 (47), 209 (40)	[169]
Coniferyl aldehyde	7.2	177.1 ([M-H]⁻), 133 (3), 134 (5), 161 (10), 162 (100), 177 (45)	[169]
km-3-O-triglucoside-7-O-glucoside	17.63	933.3 ([M-H]⁻), 429 (51), 284 (100)	[172]
Sinapoylhexose	23.4	385.1 ([M-H]⁻), 208 (32), 179 (21), 164 (100)	[172]
km-3-O-sinapoylsophoroside-7-O-glucoside	24.84	977.3 ([M-H]⁻), 429 (100), 284 (71), 285 (23)	[172]
km-3-Otrisinapoylhydroxyferuloylsophoroside	49.75	1419.4 ([M-H]⁻), 429 (100), 284 (43)	[172]
Sophoraflavonoloside	36	609 ([M-H]⁻), 447, 429, 285	[174]
Quercetin-deoxyhex	25.4	447 ([M-H]⁻), 301	[174]
Quercetin 3-*O*-sinapoyldiglucoside-7-*O*-glucoside	26.9	993 ([M-H]⁻), 831, 625, 301	[174]
Kaempferol-sinapoyl-trihexoside	28.8	977 ([M-H]⁻), 815, 653, 447	[174]
Kaempferol-3-*O*-glucoside-7-*O*-glucoside	30.1	609 ([M-H]⁻), 489, 447, 285	[174]
Isorhametin-dihexoside-*O*-glucoside	30.7	801 ([M-H]⁻), 639, 315	[174]
Isorhametin-dihexoside		639 ([M-H]⁻), 477, 315	[174]
(2R)-2-hydroxy-3-butenyl (progoitrin)	3.7	332,308,301,298,275,259,241,227,210,195	[89]

Table 3. (Continued)

Analyte (Species)	t_R [min]	Fragmentation MS/MS^3/MS^4 data (Relative intensity,%)	[Ref.]
4-Methylsuphinylbutyl (glucoraphanin)	4.0	372, 259	[89]
4-Hydroxy-3-indolylmethyl (4-hydroxyglucobrassicin)	9.7	383, 285, 275, 267, 259, 240, 176	[89]
4-Methylthiobutyl (glucoerucin)	19.4	340,291,275,259,242,227,195,178,163	[89]
4-Methoxy-3-indolylmethyl	31.2	463, 397, 299, 291, 275, 259, 235, 227	[89]
Slternariol (*Alternaria* spp.)	17.6	257 ((-)-MS-mode)	[208]
Gliotoxin (*Aspergillus fumigatus*)	13.2	263	[208]
T-2 toxin (*Fusarium culmorum* and *graminearum*)	22.5	489	[208]
Stigmasterol	0.88	343, 255, 213, 129	[144, 224]
β-Sitosterol	1.00		[144]
Fucosterol	1.06	412 M + (8)	[144]
Δ^7-Stigmastenol	1.18		[144]

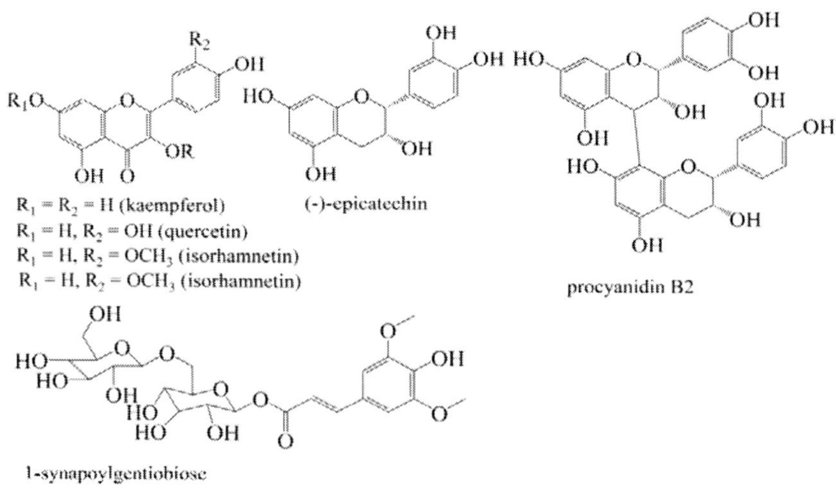

Scheme 3. (Continued).

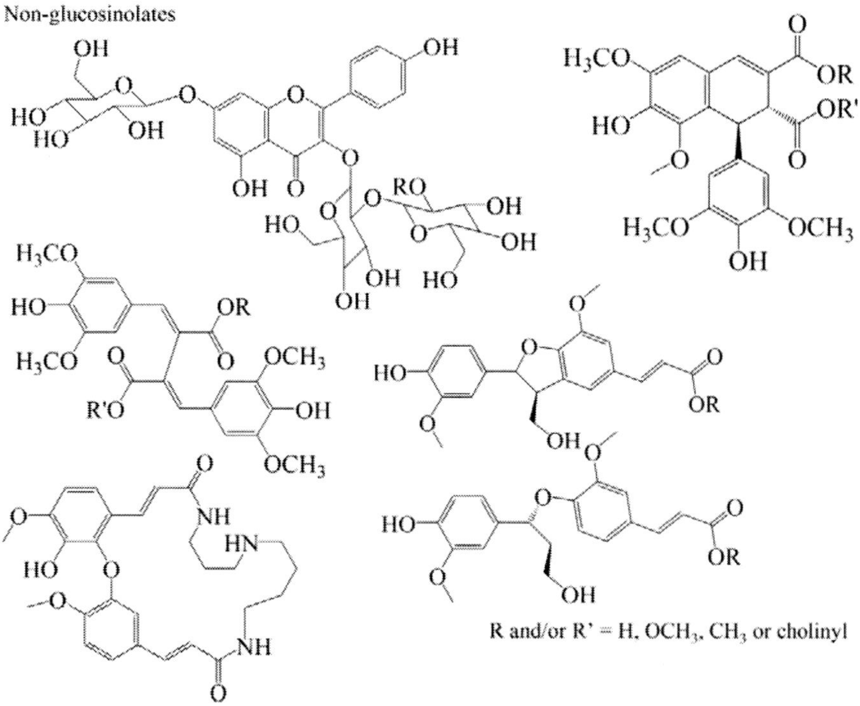

Scheme 3. Chemical diagrams of the LMW analytes in seed according [172, 173, 175].

The effect of the microwaving on the flavor characteristics of rapeseed has been examined in [176], using GC–MS. Compounds such as dimethyl di/sulfide, acetic acid, 2–methyl–1–propanol, 2/3–methyl–butanal, formate–3–methyl–1–butanol, 2,4–pentadienenitrile, methyl–pyrazine, octane, hexanal, 2–furanmethanol, fyrfural, 5–cyano–1–pentene, 2,3–dimethyl–pyrazine, 2–ethyl–3 (or 6)–methylpyrazine, 3–Ethyl–2,5–dimethylpyrazine and more have been determined. Phytoalexins and phytoanticipins (Scheme 4) are naturally produced metabolites with antimicrobial function [213–221], occurred in the rapeseed under stress y Cu^{2+}–ions [216, 219].

The MS analysis of mycotoxins has represented also a mandatory task of "Food contro," because of those secondary metabolites produced by filamentous fungi have nephrotoxic, carcinogenic, nephrotoxic or estrogenic effect [208]. Their determination in oilseed meal has been reported to [208] (Table 3).

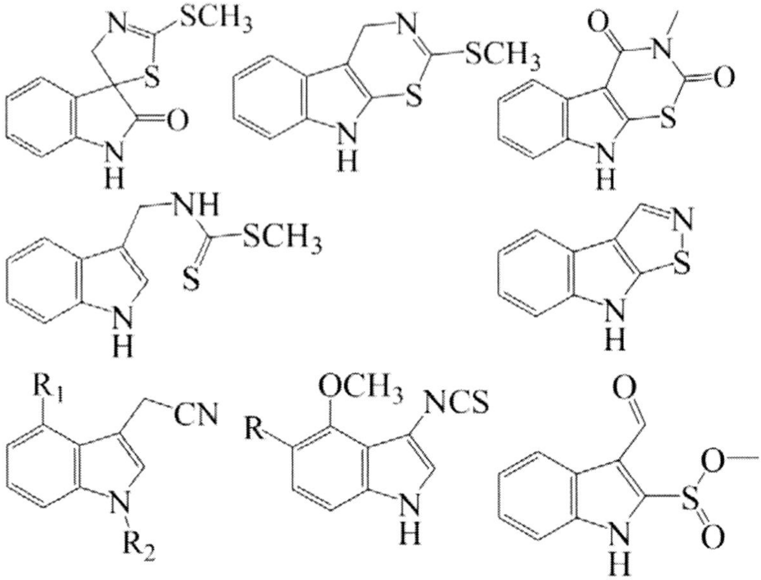

R_1=R_2=H
R_1=H, R_2=OCH_3
R_1=OCH_3, R_2=H
R=H or OH

Scheme 4. Phytoalexins and phytoanticipins in rapeseed [216].

The seed has ability to cumulate phenolic substances such as sinapine, sinapic acid (τ_R = 9.1 mins, MS/MS ($[M–H]^-$ (*m/z* 223), *m/z* 209 (Relative abundance, 6), 208 (100), 179 (36), 164 (28), 149 (2), MS^3 (MS/MS (*m/z* 208)), *m/z* 193 (8), 165 (45), 164 (100), 163 (12), 149 (24)), sinapate ester derivatives, 4-vinylsyringol (τ_R = 13.4 mins, MS/MS ($[M+H]^+$ (*m/z* 181)), *m/z* 150 (8), 149 (100), 121 (12), 93 (3)) and its trimer (τ_R = 21.6 mins, MS/MS ($[M–H]^-$ (*m/z* 539)), *m/z* 524 (100), 522 (17), 509 (8), MS^3 (MS^2 (*m/z* 524)), *m/z* 509 (25), 508 (30), 507 (100), 357 (8), 355 (6), 343 (12), MS^4 (MS^3 (*m/z* 507)), *m/z* 492 (100), 477 (52), 464 (12), 449 (21), 353 (8), 340 (9), MS^5 (MS^4 (*m/z* 492)), *m/z* 477 (100), 464 (21), 449 (28), 341 (14), 325 (9), MS^6 [MS^5 (*m/z* 477)] *m/z* 462 (100), 449 (21), 447 (16), 325 (9), 323 (18); MS^7 [MS^6 (*m/z* 462)] *m/z* 447 (100), 434 (11), 431 (10), 406 (16), 391 (12), 378 (16), 309 (18), 293 (10); MS^8 [MS^7 (*m/z* 447)] *m/z* 432 (100), 389 (77), 388 (9), 359 (21), 293 (62) as well as *cis/trans*–diastereomers of phenylindane (τ_R = 20.8 mins, MS/MS $[M–H]^-$ (359), *m/z* 345 (2), 344 (100), 329 (8), 152 (3), MS^3

[MS/MS (*m/z* 344)] *m/z* 344 (2), 329 (100), 192 (3), 177 (3), 152 (4), MS^4 [MS^3 (*m/z* 329)] 315 (3), 314 (100), 301 (10), 299 (15), 175 (9), MS^5 [MS^4 (*m/z* 314)] *m/z* 299 (100), 283 (23), 175 (31), 138 (8), 123 (7), MS^6 [MS^5 (*m/z* 299)], *m/z* 298 (19), 284 (34), 271 (66), 255 (18), 175 (100), 173 (39), 125 (58) and 123 (32), determined by LC–(or HPLC–)–ESI–MS [165, 166]. The first group compounds originate of kaempferol glycosides or from glucose and gentiobiose, respectively (Scheme 5).

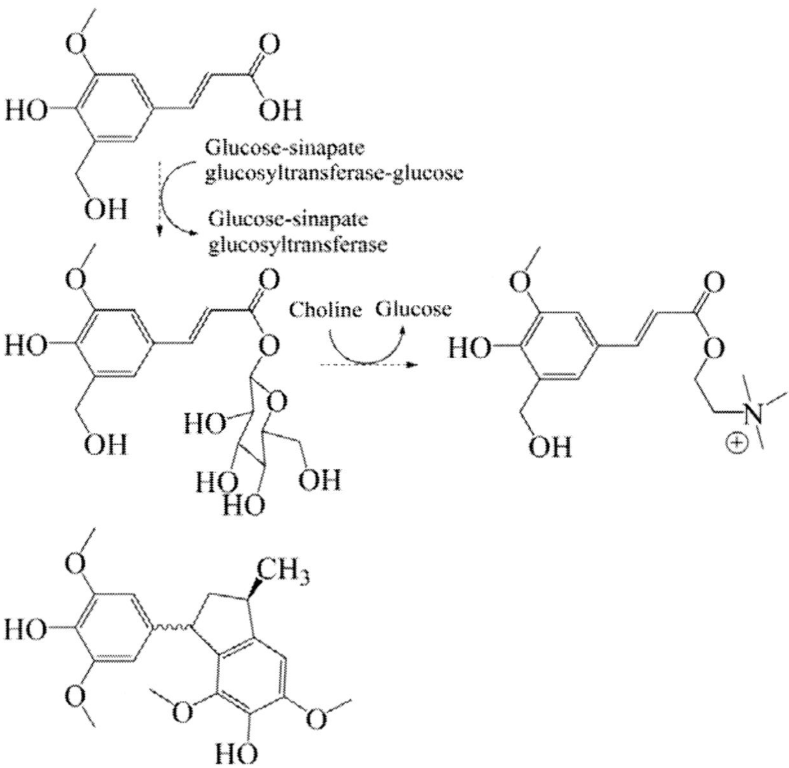

Scheme 5. Formation of sinapine in seed; Chemical diagrams of 4-vinylsyringol and *cis/trans*-diastereomers of phenylindane [165, 166].

3.3. Analysis of *Brassica Napus* Root

The plant roots have assimilated significant amount of organic substances from the rhizosphere [222]. Those compounds can be both low– and high molecular weight analytes, furthermore varying within the frame of a large dynamic range depending on physical, chemical and/or biological characteristics of the soil, the presence of metal ions such as, for example, Zn2+ or Fe2+/3+ and more. The major organic compounds in the roots represent carboxylic acids, which have been determined in the rapeseed routs studying a large scale of experimental conditions [222]. Maleate, malate, succinate, oxalate and citrate have been analyzed mass spectrometrically. To determine ultra-trace concentration limits of detection ∈ 1–47 nm by LC–ESI–MS/MS, there has been elaborated an approach involving chemical derivation of the acidic analytes, using the esterification of the molecular scaffold functionalization through formation of chlorotrimethylsilane intermediates (Scheme 6) [223].

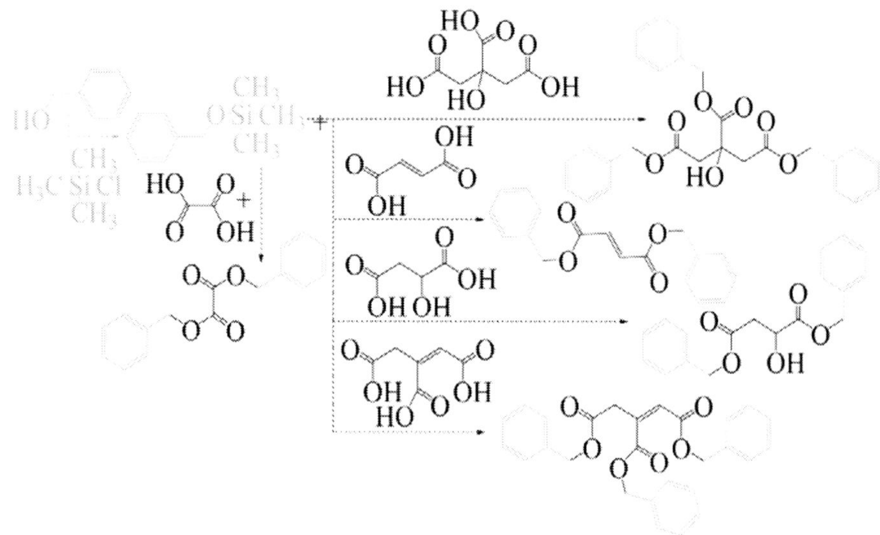

Scheme 6. Rrimethylsilane–derivative and esperification of carboxilyc acids in rapeseed roots [223].

3.4. Rapeseed Oil Analysis

3.4.1. Analysis of Naturally Occurring Products in the Rapeseed Oil

Phytosterols in the rapeseed oil we shall describe here (Scheme 7), using their classification as LMW analytes, naturally occurring in the rapeseed oil product and the rape meal. In general the phytosterols have shown high healthy effect as are widely used for nutrients. More than 200 phytosterols have been identified in the oil crops [224–227]. Cholesterol (Δ^5-cholesten-3β-ol) and stigmasterol ([24R]-24-ethyl-$\Delta^{5,22}$-eholestadien-3β-ol) are in trace concentration levels [144], while brassicasterol ([24S]-24-methyl- $\Delta^{5,22}$-cholestadien-3β-ol), campesterol ([24R]-24-methyl- Δ^5-cholesten-3b-ol), b-sitosterol ([24R]-24-ethyl-Δ^5-cholesten-3b-ol), Δ^5-avenasterol ([24Z]-24-ethylidene-Δ^7-cholesten-3b-ol) and Δ^7-stigmastenol ([24R]-24-ethyl-Δ^7-cholesten-3b-ol) are ∈ 2–58% [144]. The chromatographic and GC–MS and GLC–MS identification and quantitation is shown in Table 3 [224–227]. The APCI– and ESI–MS data have shown [226]: cholesterol (*m/z* 369 $[M+H–H_2O]^+$) (ESI/APCI), 409 ($(M+Na]^+$) (ESI); stigmasterol *m/z* 394 ($[M–H_2O]^+$) (ESI/APCI), 395 ($[M+H–H_2O]^+$) (APCI); β–sitosterol *m/z* 396 ($[M–H_2O]^+$) (ESI/APCI), 397 ($[M+H–H_2O]^+$) (APCI), 415 ($[M+H]^+$) (ESI); fucosterol *m/z* 394 ([M-H2O]+) (APCI), 395 ($[M+H–H_2O]^+$) (APCI/ESI) and 413 ($[M+H]^+$) (ESI), respectively [226].

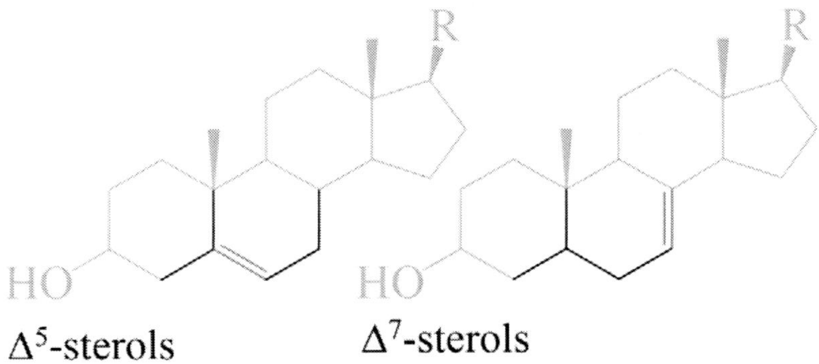

Scheme 7. Chemical diagram of Δ^7– and Δ^5–sterols [227].

Equimolar mixture of methyl 9,10-bis(methylthio)octadec-12-enoate and 12,13-bis(methylthio)octadec-9-enoat has been determined quantitatively, determining the rapeseed oil matrix effect by LC–MS/MS. Zearalenone and its derivatives (Scheme 8) have been assigned as metabolites of Fusarium fungi, colonizing grains including rapeseed [213]. These naturally occurring compounds can induce endocrine disruption by steroid receptor signaling mechanism thus altering the hormone production. For this reason they are object of food control in the edible oils. Particularly in the rapeseed oil product there is detected concentration level 40.7 $\mu g.kg^{-1}$ in a case study of marked oil [213].

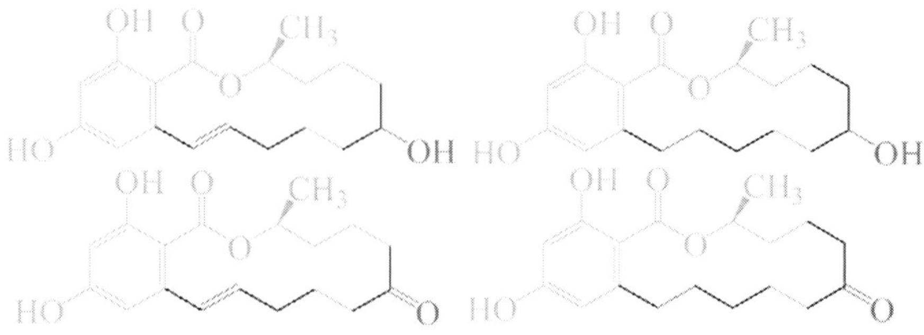

Scheme 8. Chemical diagrams of zearalenone and its derivatives [213].

Ricinoleic acid at concentration level 208 ± 40 ppm and cytokinins have been detected in rapeseed as well [213–221].

3.4.2. Analysis of Contaminating Agents in Rapeseed Oil

3.4.2.1. Analysis of Organic Pollutants

The adulteration of rapeseed oil with aniline has resulted to acute pulmonary and neuromuscular disorders in Spanish toxic oil syndrome [228–236]. From amount of 50 000–100 000 people consumed the poisoned oil the affected people are over 20000 and 350 lethal cases [230]. An immunological pathogenesis has been associated with this syndrome, causing for enormous efforts in understanding of the biochemical mechanisms governing the metabolism of the poisoned oil components in humans. As proposed causative agents are denoted N-(2-hydroxy-3-butenyl)-N'-phenylthiourea and its cyclic product 1-phenyl-5-vinyl-2-imidazolidinethio, determined in oil case samples by GC–EI–MS [230]. The fragmentation pathways are shown in Scheme 9. To the toxic oil syndrome has also been associated contamination with pentachlorophenol [234].

Toxic oil syndrome has also been associated with anylines 3-(N-phenylamino)-1,2 propanediol, [M+H]+ (m/z 168), 2-hydroxy-3-(N-phenylamino)propyl linoleate (m/z 430), 2-hydroxy-3-(N-phenylamino)propyl oleate (m/z 432), 2-hydroxy-3-(N-phenylamino)propyl heptadecanoate (m/z 420), 2-linoleyloxy-3-(N-phenylamino)propyl linoleate (m/z 692), 2-oleyloxy-3-(N-phenylamino)propyl linoleate (m/z 694), 2-linoleyloxy-3-(N-phenylamino)propyl oleate (m/z 694), 2-oleyloxy-3-(N-phenylamino)propyl oleate (m/z 696), (OO(1,2)PAP), 2-linoleyloxy-3-(N-phenylamino)propyl heptadecanoate (m/z 682), 2-oleyloxy-3-(N-phenylamino)propyl heptadecanoate (m/z 684), 2-heptadecanoyloxy-3-(N-phenylamino)propyl (m/z 672), N-phenyl linoleamide (m/z 356), N-phenyl linolenamide (m/z 354), N-phenyl oleamide (m/z 358), N-phenyl eicosanamide (m/z 386), N-phenyl heptadecanamide (m/z 346) [159]. An analytical protocol for determination of polycyclic hydrocarbon contaminating organics in rapeseed oil by DART–MS has been reported to [214]. An object of strict control in rapeseed oil food product is also 3-chloro-1,2-propanediol a food–process contaminating agent, having carcinogenic effect *in vitro* [187, 209, 215, 219]. The maximum admissible daily concentration level 2 $\mu g.kg^{-1}$ body weight day-1 has been determined by Joint Food and Agriculture Organization of the United Nations/World Health Organization Expert Committee on Food Additives in 2007 [215]. In addition to the quantitation of organic contaminating agents causing toxic oil syndrome, in this section of the chapter we shall describe among the most important tasks relating the food control, particularly the rapeseed oil analysis, i.e., the quantitation of pesticide residues in oil foodstuffs, having great impact for the human health risk along with above shown contamination type [237–240]. Like all agricultural crop protecting agents, the employment of fungicides, herbicides and pesticides in rape crop prevention can causes to remaining of those substances and/or their metabolites as well as degradation products at the harvest stage and thus to contaminate both the final oil product and the meal. The maximal residue levels of pesticides for rapeseed products, fixed by Regulation Directive 396/2005 of the European Commission are listed in Tables 5 and 6 [241]. The same Tables summarize MS data, employed in the analytical protocol for rapeseed oil analysis are reported in [237–241]. Firstly in the literature the analysis of naturally occurring pesticides of rotenone type in rapeseed oil products has been reported in [242], employing the MALDI–MS method. Due to their high toxicity to humans causing for neurodegenerative atypical parkinsonian syndrome, there is strict regulation for the European Union Countries limiting the maximal residue levels to 40 ppb [242–244]. Along with the achieved concentration limits of detection 0.14 $ng.g^{-1}$, recoveries $\in$ 99.98–99.36% (SD(yEr±) $\in$ 0.00156 – 0.00575) for levels of pollutant 1 of 1.00, 5.0 and 25 μg and $\in$ 99.968–98.772% (SD(yEr±) $\in$ 0.00415–0.12422) for levels of pollutant 2 of 1.00, 5.0, and 25 ng in spiked oil sample sets, there has been presented a methodological contribution for sample preparation techniques applied for MALDI–MS measurements, where the classical polyfunctional carboxylic acids have been replaced with a new designed set of crystals of organic salts (Figure 3), which optical spectroscopic characteristics make them suitable matrix components both for UV–and IR–MALDI mass spectrometry [75, 242].

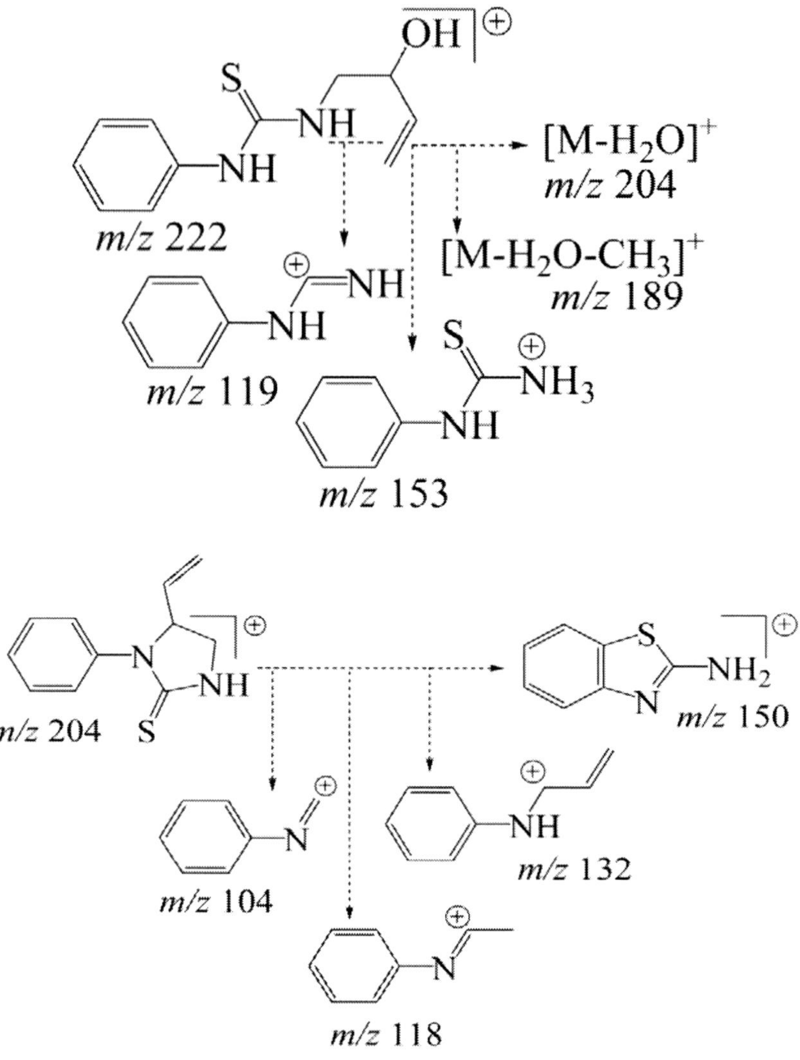

Scheme 9. EI–MS fragmentation of N–(2–hydroxy–3–butenyl)–N'–phenylthiourea and 1–phenyl–5–vinyl–2–imidazolidinethio [230].

Table 4. Maximum residue levels [mg.kg^{-1}] of pesticides in rapeseed according European Union Directive 396/2005 (2011) [243–246]; MS/MS fragmentation ions (abundance, %)

Pesticide	Level	Precursor MS ion	MS ion (abund.%)
Acetamiprid	0.01	223	126 (100)
Azoxystrobin	0.5	404	372 (100)
Chlorfenvinphos	0.05	359	155 (100)
Clomazone	0.02	240	125 (100)
Difenoconazole	0.05	251	251 (100)
Fluorochloridone	0.1	292	292 (100)
Metalaxy-M	0.1	280	192 (71)
Oxadiazon	0.05	345	220 (100)
Prochloraz	0.5	376	308 (100)
Thiacloprid	0.3	253	126 (100)
Trichlorfon	0.1	257	109 (100)
Atrazine	0.05	216	174 (100)
Carboxin	0.1	236	143 (100)
Clethodim	1	360	164 (100)
Dichlorvos	0.01	221	109 (100)
Fluazifop-P-butyl	0.15	384	328 (63)
Imidacloprid	0.1	256	209 (100)
Metolachlor	0.1	284	252 (100)
Phoxim	0.02	229	129 (100)
Tebuconazole	0.5	308	70 (100)
Triadimefon	0.2	294	197 (100)
Acetochlor	0.2	270	224 (100)
Carbendazim	0.1	192	132 (12)
Chlorpyrifos	0.05	350	198 (100)
Diazinon	0.02	305	169 (100)
Diniconazole	0.05	324	70 (100)
Haloxyfop-P-methyl	0.2	376	316 (100)
Methidathion	0.05	303	145 (100)
Paclobutrazol	0.02	294	70 (100)
Propiconazole	0.1		
Thiamethoxam	0.05	292	211 (100)

Table 5. Negative-ESI precursor ion mass spectra of authentic gibberellin in rapeseed [200]

	$[M–H]^-$	*M/z* fragment
GA8	363	321(40), 275(100), 257(45),119(55)
GA1	347	303(50), 259(100)
GA6	345	301(35),257(80),239(40),161(40), 119(100)
GA20	331	287(100), 250(10),225(10),219(15),173(50), 147(30)
GA13	377	359(100), 303(45),259(5),215(15)
GA19	361	317(70), 273(100), 255(45),229(45),203(60), 133(40)
GA34	347	303(25), 259(100), 241(35),199(20)
GA51	351	287(100), 243(75),219(60),182(20)
GA7	329	287(5), 223(100), 211(15),168(10)
GA4	331	313(50), 287(100), 257(75)
GA53	347	329(100), 303(75),233(35),189(70)
GA24	345	301(90), 257(100), 213(60),135(35)

Table 6. Relative peak area of aldehydes from condensates of rapeseed oil fumes

	190-200	230-240	270-280	M/z	
3-Methyl-2-butenal	0.78	0.50	0.22	55,41,83,39,69	[161]
Hexanal	2.09	2.49	2.73	44,56,41,43,57,100	[161]
2-Hexenal	0.38	0.47	0.25	41,55,69,83,39,98	[161]
Heptanal	0.24	0.54	1.54	43,41,70,44, 56,114	[161]
2-Heptenal	3.30	4.23	4.42	41,83,55,57,56,112	[161]
2,4-Heptadienal	8.11	4.58	1.19	81,39,41,53,67,95	[161]
Octanal			1.05	43,41,57,55,84,69	[161]
2-Octenal		0.46	0.90	41,55,70,39,83,126	[161]
Nonanal		2.27	7.47	41,43,57,56,44,142	[161]
2-Nonenal			0.72	41, 55,70,83,96,140	[161]
Conc, T 240-280°C					
1,3-Butadiene				504 $ng.L^{-1}$	[162]
Benzene				2391 $ng.L^{-1}$	[162]
Acrolein				391.8 $\mu g.L^{-1}$	[162]
2,4-nonandienal		0.47	0.27	81,41,39,53,138	[161]
decanal			0.28	43,41,57, 44,70,71,156	[161]
2-decenal		1.99	7.42	41,43,55,70,39,83, 154	[161]
2,4-decadienal		2.01	2.98	81,41,39,53,138	[161]
Undecanal				43,41,57,55,67,170	[161]

	190-200	230-240	270-280	M/z	
2-undecenal		1.89	7.84	41,70,55,43,83,168	[161]
2,4-undecadienal			0.21	81,41,55,39,166	[161]
dodecanal				43,57,41,82,68,184	[161]
9-octadecenal			0.23	55, 69,43,67,83,266	[161]
9,17-octadecadienal			0.30	67,81,41, 95,53,264	[161]
Conc, 240-280°C					
Formaldehyde				71.2 μg.L^{-1}	[162]
Acetaldehyde				306.9 μg.L^{-1}	[162]

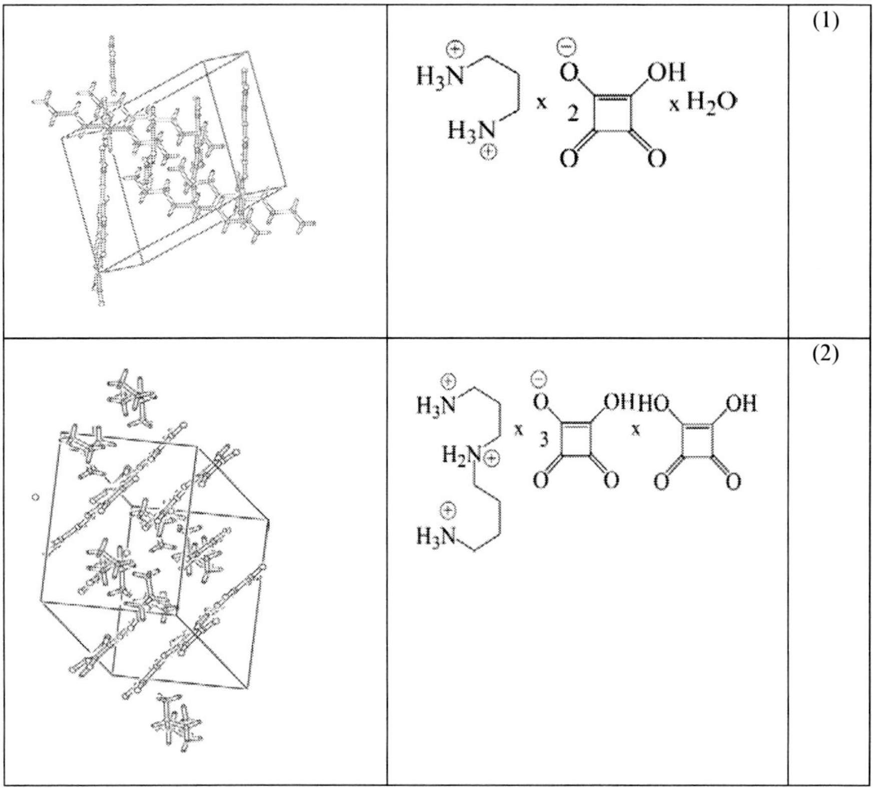

Figure 3. Crystal structures of N[1]-(3-amino-propyl)-butane-1,4-diammonium tris(hydrogensquarate) squaric acid (1) and pentane-1,5-diammonium bis(hydrogensquarate) monohydrate (2) [242]; Chemical diagrams.

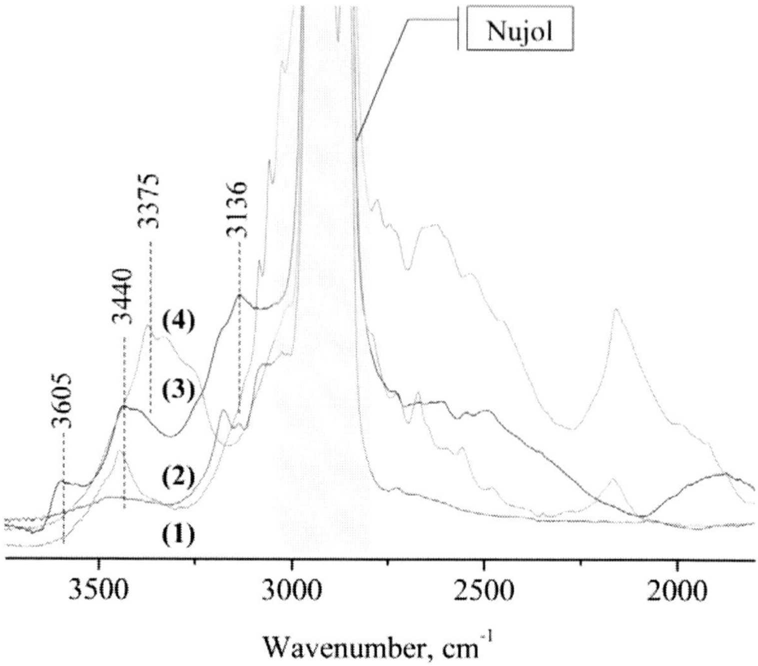

Figure 4. 3750–1800 cm^{-1} solid–state IR–spectra of N^1-(3-amino-propyl)-butane-1,4-diammonium tris(hydrogensquarate) squaric acid (1), pentane-1,5-diammonium bis(hydrogensquarate) monohydrate (2), propane-1,3-diammonium bis(benzene-1,2,4-tricarboxylate) dihydrate (3) and pentane-1,5-diammonium squarate (4) measured, employing the nujol–mull technique.

Among the crucial factor in the molecular design of new matrix components for MALDI–MS sample preparation are the optical properties of the crystals, where when the analysis has involved desorption/iuonizations laser irradiation source within the UV – region, than the matric crystal should have properties close to the exitation wavelength. By contrast, for the needs of the IR–MALDI–MS it is need absorption of crystals ∈ 3300–3000 cm^{-1}, where mainly ν_{OH} and ν_{NH} stretching vibrations of NH– or OH–groups have occurred. Given that the implemented new crystals for rapeseed oil analysis by MALDI–MS [242] have shown absorption UV–bands ∈ 200–350 nm of n→π* transitions as well as IR–bands ∈ 3300–3000 cm^{-1} of ν_{OH}, ν^{as}_{N+H3}, $\nu^{as}_{N+'H3}$ and ν^{s}_{NH3}, respectively (Figures 4 and 5).

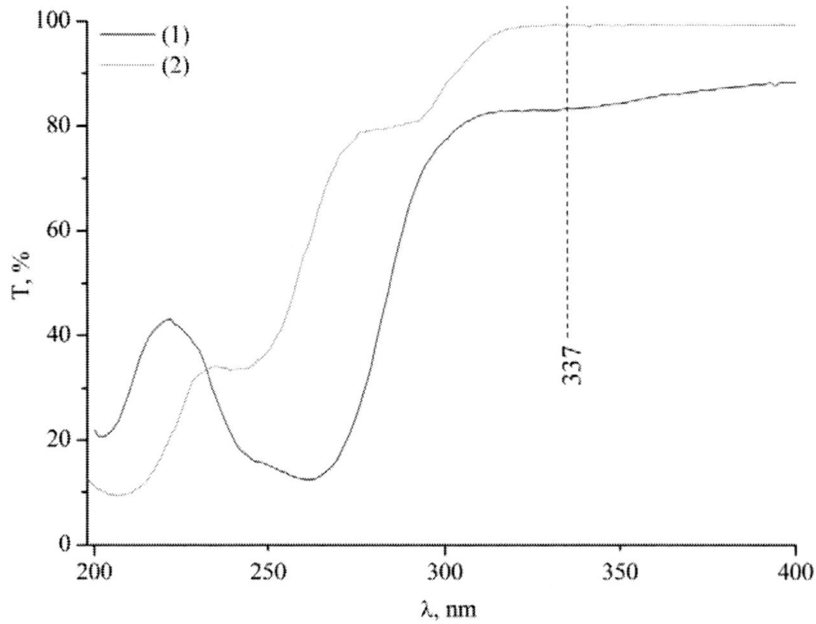

Figure 5. UV–spectra pentane-1,5-diammonium bis(hydrogensquarate) monohydrate (1), propane-1,3-diammonium bis(benzene-1,2,4-tricarboxylate) dihydrate (2) [242].

As it has been broadly discussed [75], there has clear trend in search of new matrix component, replacing the neutral crystals and co–crystals for MALDI–MS sample preparation, because of the salts of organic salts have advantages versus the neutral polyfunctional banzoic acid derivatives consisting in regular polycrystalline growth increasing the sample – to – sample reproducibility and spot–to–spot repeatability of measurements, thus increasing the method performance parameters in addition to the indicating of successful ionization of large scale LMW analytes in various environmental and foodstuffs matrices [75].

3.4.2.2. Analysis of Inorganic Pollutants

In general the determination of trace elements is an important task in assessment of the quality of edible oils involving the freshness and the period of the storage [247–251]. The presence of the heavy metal ions in the final oil products has been associated with their migration from the soil to oilseeds. Despite that some transition metals are essential for the humans like the described below Zn–complexes, environmental exposure and contamination from them can cause for a bioaccumulation both in the oil and in the consumers, where with increasing in concentration a toxic effect rises [248]. The analysis of the heavy metal content has mandatory involved the series of the well known toxic heavy metals and their organometallic As, Hg, Cr, Cd, and Pb, respectively [247]. Among the sensible instrumental analytical methods for analysis of heavy metals in foods, in general, and in parallel a mass spectrometric based approaches is ICP–MS [75]. In spite of the known capacity of this method for trace concentration level analysis, it application is accompanied with series sample pretreatments, which in the case of the oil analysis are complicated further due to the viscosity of the oil foodstuff matrix. In order to overcome those drawbacks there are developed analytical approaches for extraction of the heavy metals from the oil sample using the EDTA as one, becoming universal ligand [247]. Concentration ranges 75.7±3.6 $\mu g.kg^{-1}$ (Mg), 12.5±1.8 $\mu g.kg^{-1}$ (Al), 0.29±0.04 $\mu g.kg^{-1}$ (Cu), 25.2±4.1 $\mu g.kg^{-1}$ (Zn) and 0.02±0.008 $\mu g.kg^{-1}$ (Cd) in rapeseed oil have been reported [247]. But soft–ionization MS methods have provided crucial information about the complexes of essential metal ions in the rapeseed plant tissue [249]. Thus, those results have highlighted the promising prospectives of last MS approaches for metallomics [75]. The migration of thallium ($Tl^{I/III}$) from soil to rapeseed plant has been examined by ICP–MS [251], governing the study by the fact that this $Tl^{I/III}$ is not essential metal and thus it is toxic for the human health. Accounting for the fact that the ionic radii of Tl+– and K+– ions are similar it has been distributed in the plant cells, similarly. Tl^{+}–ion is accumulated intra-cellularly in addition to a proposed enzymatic inhibition effect like Hg^{2+}– and Pb^{2+}–containing species. In general, high accumulation effect has occurred [251].

3.4.3. Chemical Analysis of Reactions and Products of Thermal Treatment of Rapeseed Oil

The major focus of this section of the chapter is on various chemical reactions and products occurred under thermal treatment of rapeseed oil, describing known auto– and thermal oxidation processes, polymerization and hydrolysis as well as cyclic and fission reactions, comprehensively elucidated employing methods of the mass spectrometry, based on soft–ionization techniques, largely employing coupled instrumental schemes by GC–, LC– or HPLC– as preliminary separation steps [75, 203, 252–260]. The chemical analysis carried out, has involved determination of the reaction products of rapeseed oil under different experimental conditions encompassing temperature ranges ∈ 190–200, 230–240 and 270–280°C [161, 162]. The experimental designs have been based on popular ways for cooking worldwide, furthermore involving investigation of the processes in the thermally treated oil itself and the corresponding fumes content. The importance of the results, which have been achieved has been associated with numerous studies on the physiological effect of chemicals resulted from the thermal treatment of rapeseed oil as well as volatile compounds, which are in their major part aldehyde acids and aldehyde, in addition to simple unsaturated hydrocarbons such as, for example, 1,3–butadiene and benzene (Table 6). Among various aldehyde there has extensively been studied 4–hydroxy nonenal [32] obtained under deep–fat frying within a temperature range 240–280°C, showing cytotoxic, hepatotoxic, mutagenic and genotoxic effect. The 9–oxononanoic acid, which is major aldehyde acid as a result of oxidation of lipids under thermal treatment at $T > 200^{\circ}C$ can induce hepatic lipid peroxidation and thus to affect the hepatic metabolism. Furthermore, in general, aldehyde have been described as high chemically reactive compounds having ability to bond to free amino groups in large set of essential amino acids, thus forming aldimines or to take part in coupling processes with sulfhydryl groups. Same is valid for the lipid peroxides, where numerous in vivo studies have shown cytotoxic, mutagenic and teratogenic effect [158]. Despite fact that those products have concentration levels $\mu g.g^{-1}$ in the foods, there has rather lack of knowledge about their physiological effect. Accounting for the fact that the major components in the oils are fatty acids, both saturated and unsaturated as well as that rapeseed oil has shown a high level of unsaturation, which one hand make it a healthy food product. On the other hand the presence of high concentration of allyl groups has resulted to a high sensibility of the oil towards namely oxidation reaction and radical reactions. The formed autoxidation products have mainly downed into volatile products with concentrations in the oil lower than 0.0015 $\mu g.L^{-1}$ (2,6-*cis*/*trans*-nonadienal) or 0.0001 $\mu g.L^{-1}$ (1-octene-3-one) [32, 163] (Scheme 10). Nevertheless that part of the formed aldehyde remains in the oil, as it was shown above, furthermore they are bonded to the lipids through ester functional groups. These bonded aldehyde adsorbed into the frying food can have negative healthy effect for the consumers. For these reasons the development of accurate and precise protocols for determination of both volatile and bonded reaction products of the thermal treatment of oil has become a primary task for the field of the "Food control," evaluating he risk for the human health [32, 143, 158, 161, 163, 255–258].

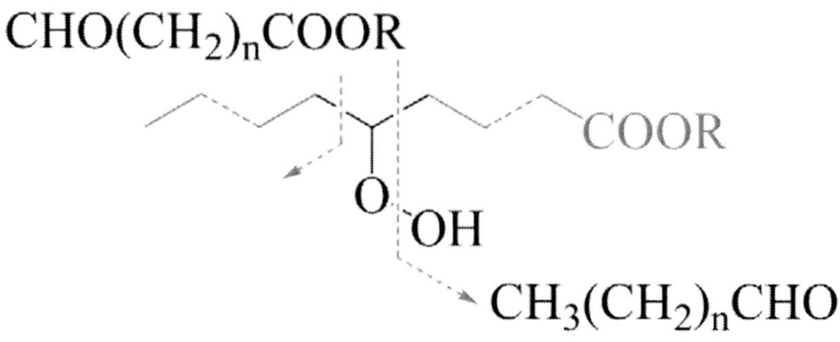

Scheme 10. β-scission of fatty acids hydroperoxide esters [32].

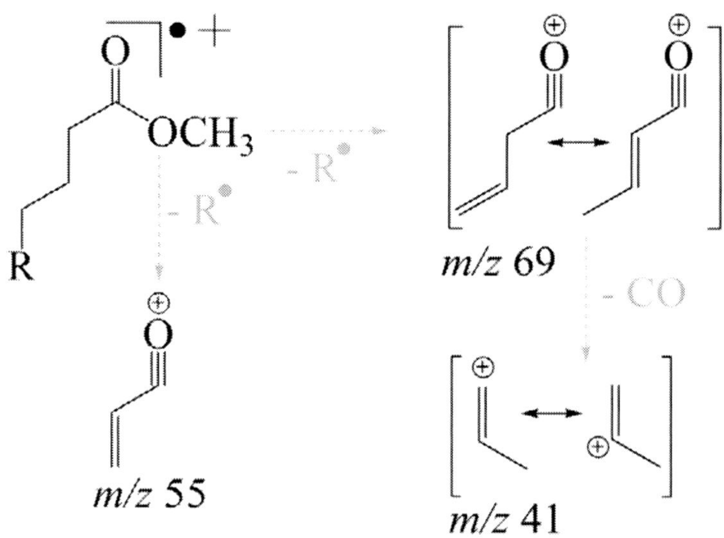

Scheme 11. MS fragmentation of saturated fatty acids esters [33].

The analysis of volatile compounds has usually involved GC–MS approach, providing meaningful information for the contain of fumes from frying of rapeseed oils. Significantly more complicated is the determination of the bonded aldehyde in the oil, where even a qualitative analysis is a difficult research task, due to hundreds possible combinations of bonded products between the fatty acids and their oxidized derivatives. In this respect there have been reported series developments based on preliminary TLC combined with HPLC–MS or employment of HPSEC for separation of polar from the non-polar fraction [32]. The 100– and 500–Å *gel columns p*acked with styrene–divinylbenzene copolymer have been utilized (Scheme 10).

The MS fragmentation of the methyl esters of fatty acids has exhibits peaks at m/z 87, 74 and 59 [CH O-C(=OH$^+$)-CH$^{\bullet}_2$] and [CH OC=O]$^+$. The major peaks for the unsaturated fatty acids are at *m/z* 41, 55, 69, however, they can be characteristics for saturated derivatives as well as via the shown fragmentation Scheme 11 [32, 33]

Direct analysis of major TGs by transmission-mode DART mass spectrometry and HPLC (or ultra–HPLC)–MS of the thermally treated rapeseed oils has been reported as well [203–205].

The pan-frying at 160°C has resulted to oxidation of sitosterol/sitostanol as well as polymerization of the lipid mass in the rapeseed oil (Figure 6) [206].

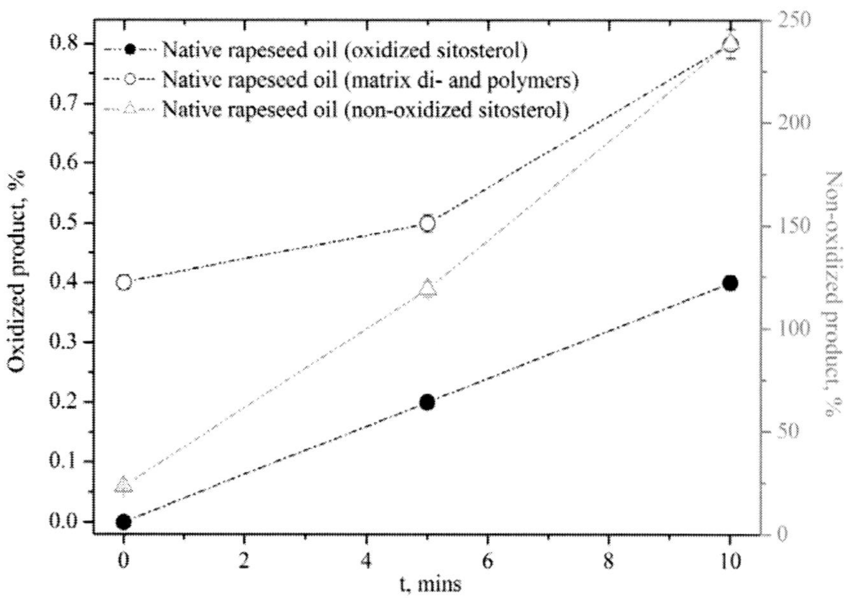

Figure 6. Oxidation of sitosterol and lipid matrix in native rapeseed oil [206].

The quantitation of canolol content as a function of the roasting time at different temperatures has been studied in (Figure 7) [207].

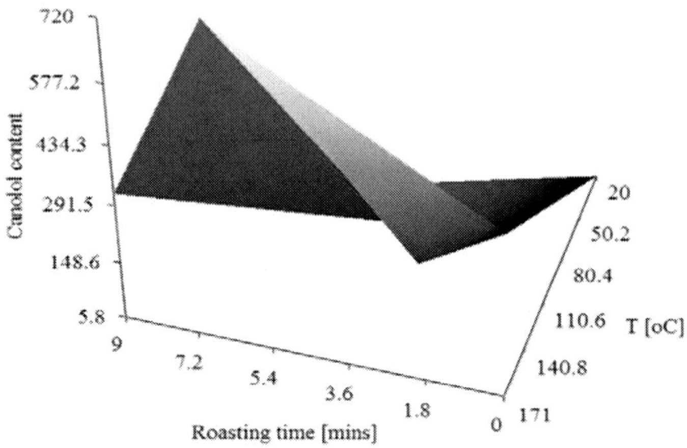

Figure 7. Roasting effect on the concentration of conolol in rapeseed oil [207].

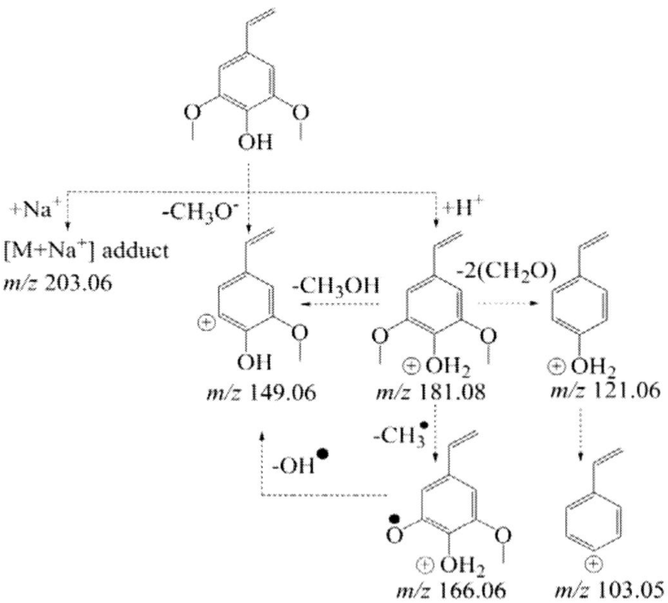

Scheme 12. Fragmentation pattern of conolol [210].

The MS spectrometric profile of conolol, involved into it identification and determination quantitatively has comprehensively been described (Scheme 12) [210].

3.5. Rapeseed Melt Analysis

Along with beneficial nutritional properties of the rapeseed oil itself, the usage of the rapeseed meal after oil extraction as feedstock for animals has met limitation, due to the high content of glucosinolates (Scheme 13) [199]. It has been proposed that glucosinolates are risk agents promoting cancer [88, 261, 262]. There have been known about 100 different substituted glucosinolates, from which ones four major glucosinolate components in the rapeseed meal occurred, i.e., glucobrassicin (R_1 = R_2 = H), 4–hydroxy–glucobrassicin (R_1 = OH, R_2 = H), 4–methoxyfglucobrassicin (R_1 = OCH_3, R_2 = H) and neoglucobrassicin (R_1 = H, R_2 = OCH_3), respectively (Scheme 13 [147], in addition to known olefinic–, methylsulfinylalkyl– and aryl–glycosinolates [88, 90]. During a seeding process they are bocked down under an enzymatic reaction with participation of myrosinase E.C.3.2.3.1 yielding to isothiocyanates, thiocyanates, nitriles and oxasolidinethiones (Schemes 14 and 15). These compounds can be harmful for the animals [147, 148, 157]. On the other hand the meal contains significant amount of essential amino acids, which are desirable for the human food supplements and animal feeds [165]. In this respect, enormous effort has been devoted to overcome this, implementing in the agricultural practice of a new variety of rapeseed, namely so–called "double zero" (0,0) crop in the agricultural practice, showing a significant reduction of the discussed glucosinolates, for example the concentration limits of detection of sinigrin 2.85 ng has been detected by LC–APCI–MS [157]. Despite this successes in improvement of nutrition properties of rapeseed meal, the analysis of glucosinolates remains as a task for control of the foodstuffs quality, so that, the MS fragmentation of various analytes has comprehensively been studied by HPLC–ESI–$(MS)^n$ (Table 3) [89]. A minor product, O-α-D-galactopyranosyl-(l→6)-O-α-D-galactopyranosyl-(1→lL)-myo-inositol (Scheme 16) of biosynthesis of stachyose, which is the major oligosaccharide in rapeseed meal has been identified in [149], along with acetates of 2,3,4,6-tetra-O-methyl-D-galactose, 2,3,5-trl-O-methyl-L-arabmose, 2,3,4-tri-O-methyl-D-glucuromc acid, 2,4,6-tn-O-methyl-D-galactose, 2,3,4-tn-O-methyl-D-galactose, 2,3-di-O-methyl-L-arabmose, 2,3-

di-O-methyl-D-galactose, 3-O-methyl-L-arabmose and 2-O-methyl-D-galactose [150], respectively.

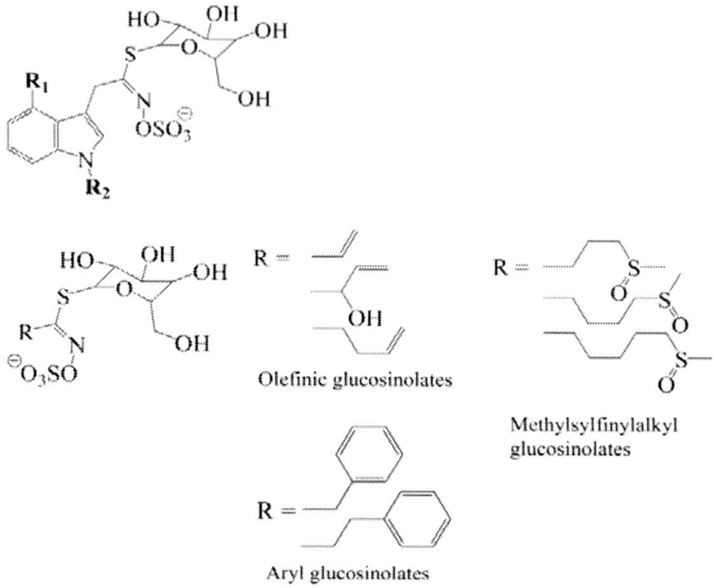

Scheme 13. Major glucosinolates in rapeseed meal [88, 90, 147].

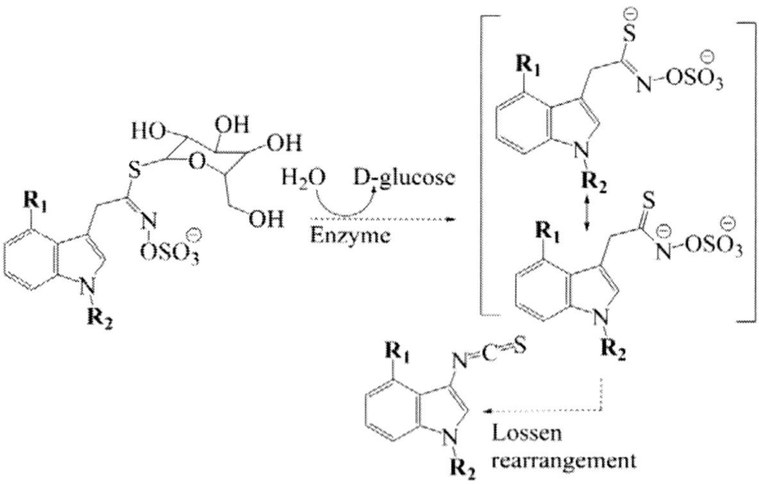

Scheme 14. Breakdown of glucosinolates [147].

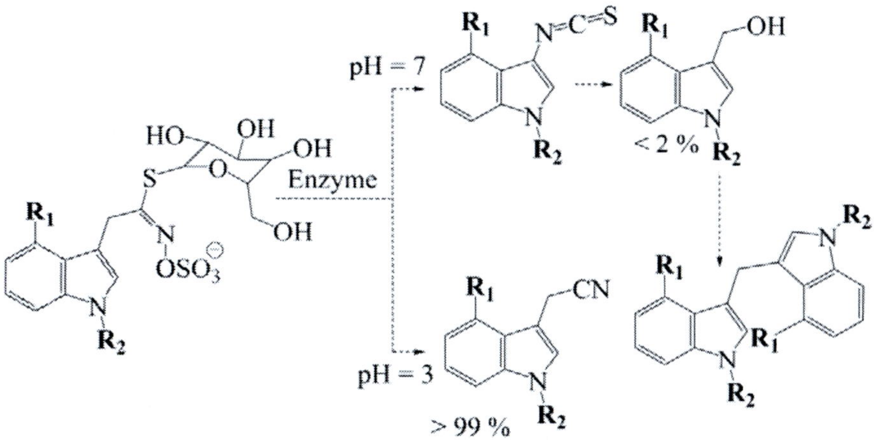

Scheme 15. Enzymic hydrolysis of glucosinolates [147].

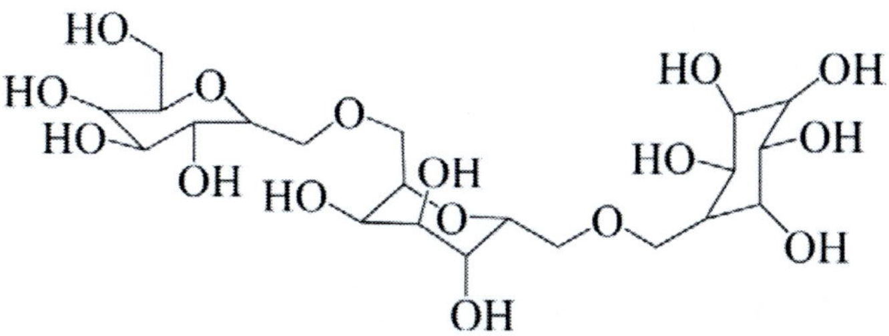

Scheme 16. Chemical diagram of O-α-D-galactopyranosyl-(l→6)-U-a-D-galactopyranosyl-(1→lL)-myo-inositol [149].

Another drawback to the wide gene non manipulated application of rapeseed meal for animal feed originates from the high content of polysaccharides containing arabinose, xylose, manose galactose and uronic acid as residues in the cell walls, which cannot be degraded by endogenous enzymes of mono-gastric animals [263–266]. Moreover the structures of those polysaccharides are still not well understood, thus causing for a pretreatment of the rapeseed crop in order to improve the nutrition properties for animal consumption.

Conclusion

In this short chapter, we wanted to introduce the readers to great instrumental features of matrix–assisted laser desorption ionization mass spectrometry for analysis of Brassica napus plant as a whole and, particularly, it edible foodstuff product having beneficial nutritional properties for human daily life, i.e., the rapeseed oil. Shortly we have considered known contributions of MALDI–MS method to qualitative and quantitative analysis of macromolecules in rapeseed oil. But, in this chapter the special concern is about low molecular weight analyte determination. Thus we refer broadly the prospectives of the method as an analytical instrumental approach in the field of "Food control" encompassing application on a large scale of analytes, involving products of thermal treatment of rapeseed oil, organic and inorganic pollutants as well as multicomponent complex analysis of flowers and hulls, seed and roots, employing its ultra high resolving power allowing quantitation in complex homogeneous and heterogeneous mixtures at fmol–to–attomol concentration limits of detection, along with it flexible sample preparation techniques and ability for imaging and direct assay.

References

[1] Chen, X., Yuan, Y., Lai, Y., Wang, L. Impact of Composition and Molecular Structure Upon the Cold Flow Properties for Biodiesel, *Acta Petrol. Sin.* 2009, *25*, 673–677.

[2] Chen, X., Lu, C., Lai, Y., Ge, W., Li, S., Tao, J. Improving Flow Properties of Rapeseed-Based Biodiesel at Low Temperatures, 1st Int. Conf. Energy Environm. Protec. ICEEP 2012, Hohhot, China, 23–24/06/2012, Code 89966, *Adv. Mater. Res.* 2012, *512–515*, 426–431.

[3] Zeman, A., Biodegradable Lubricants – Analysis of Metalworking Oils and Hydraulic Fluids by Mass Spectrescopy, *J. Synth. Lubric.* 1996, *13*, 149–161.

[4] Zeman, A., Ebenhoch, M., Muller, M. Biodegradable hydraulic fluids – Formation of dimers and oligo-/polymers during thermo-oxidative ageing, *Tribol. Schmierungstechn.* 1998, *45*, 4–9.

[5] Kulkarni, M., Dalai, A., Bakhshi, N. Utilization of green seed canola oil for biodiesel production, *J. Chem. Technol. Biotechnol.* 2006, *81*, 1886–1893.

[6] Qi, T., Lu, H., Jiang, W., Liang, B. The structure of Zn/Al Complex Oxides and Their Catalytic Activity to the Transesterification in Biodiesel Production, *Chem. React. Engin. Technol.* 2009, *25*, 52–58.

[7] Paredes, E., Can, S., Quetel, C. Examining the Possibility of Using Ratios of Element Mass Fractions as Markers for Origin Determination of Soya and Canola Based Biodiesels. Measurement Procedure Validation to Underpin this Study, *Fuel* 2014, *123*, 248–255.

[8] Kwon, E., Yi, H., Jeon, Y. Boosting the Value of Biodiesel Byproduct by the Non-catalytic Transesterification of Dimethyl Carbonate via a Continuous Flow System Under Ambient Pressure, *Chemosphere* 2014, *113*, 87–92.

[9] Norouzi, S., Eslami, F., Wyszynski, M., Tsolakis, A. Corrosion Effects of RME in Blends with ULSD on Aluminium and Copper, *Fuel Proc. Technol.* 2012, *104,* 204–210.

[10] Milne, T., Evans, R., Nagle, N. Catalytic Conversion of Microalgae and Vegetable Oils to Premium Gasoline, with Shape-Selective Zeolites, *Biomass* 1990, *21*, 219–232.

[11] Holcapek, M., Jandera, P., Fischer, J., Prokes, B. Analytical Monitoring of the Production of Biodiesel by Highperformance Liquid Chromatography with Various Detection Methods, *J. Chromatogr. A* 1999, *858*, 13–31.

[12] Pedersen, J., Lngemarsson, A., Olsson, J. Oxidation of Rapeseed Oil, Rapeseed Methylester (REM) and Diesel Fuel Studied with GC/MS, *Chemosphere* 1999, *38*, 2467–2474.

[13] Fabritius, D., Schaefer, H., Steinbuechel, A. Bioconversion of Sunfower Oil, Rapeseed Oil and Ricinoleic Acid by Candida Tropicalis M25, *Appl. Microbiol. Biotechnol.* 1998, *50*, 573–578.

[14] Fehling, E. Analysis of Estolides in Technical Hydroxylated Fatty Acids from Plant Oils, *J. Am. Oil Chem. Soc.* 1995, *72*, 355–359.

[15] Ratsameepakai, W., Herniman, J., Jenkins, T., Langley, G. Evaluation of Ultrahigh-Performance Supercritical Fluid Chromatography–Mass Spectrometry as an Alternative Approach for the Analysis of Fatty Acid Methyl Esters in Aviation Turbine Fuel, *Energy Fuels* 2015, *29*, 2485–2492.

[16] Chyhyrynets, O., Vorob'iova, V. Anticorrosion Properties of the Extract of Rapeseed Cake as a Volatile Inhibitor of the Atmospheric Corrosion of Steel, *Mater. Sci.* 2013, *49*, 318–325.

[17] Pyla, S., Schietekata, C., Van Geema, K., Reyniers, M., Vercammen, M., Beens, J., Marin, G. Rapeseed Oil Methyl Ester Pyrolysis: On-line

Product Analysis Using Comprehensive Two-dimensional Gas Chromatography, *J. Chromatogr. A* 2011, *1218*, 3217–3223.

[18] Samanya, J., Hornung, A., Apfelbacher, A., Vale, P. Characteristics of the Upper Phase of Bio-oil Obtained from Co-pyrolysis of Sewage Sludge with Wood, Rapeseed and Straw, *J. Anal. Appl. Pyrol.* 2012, *94*, 120–125.

[19] Briones, R., Serrano, L., Llano-Ponte, R., Labidi, J. Polyols Obtained from Solvolysis Liquefaction of Biodiesel Production Solid Residues, *Chem. Engin. J.* 2011, *175*, 169–175.

[20] Eide, I., Neverdal, G., Westad, F. Detection of 5 ppm Fatty Acid Methyl Ester (FAME) in Jet Fuel Using Electrospray Ionization Mass Spectrometry and Chemometrics, *Energy Fuels* 2010, *24*, 3661–3664.

[21] Kang, S., Li, X., Li, B., Fan, J., Chang, J. Effects of Lignins on Antioxidant Biodiesel Production in Supercritical Methanol, *Energy Fuels* 2011, *25*, 2746–2748.

[22] Cecrle, E., Depcik, C., Duncan, A., Guo, J., Mangus, M., Peltier, E., Stagg-Williams, S., Zhong, Y. Investigation of the Effects of Biodiesel Feedstock on the Performance and Emissions of a Single-Cylinder Diesel Engine, *Energy Fuels* 2012, *26*, 2331–2341.

[23] Flitsch, S., Neu, P., Schober, S., Kienzl, N., Ullmann, J., Mittelbach, M. Quantitation of Aging Products Formed in Biodiesel during the Rancimat Accelerated Oxidation Test, *Energy Fuels* 2014, *28*, 5849–5856.

[24] Elghawi, U., Theinnoia, K., Sitshebo, S., Tsolakis, A., Wyszynski, M., Xua, H., Cracknell, R., Clark, R., Mayouf, A. GC–MS Determination of Low Hydrocarbon Species (C1–C6) from a Diesel Partial Oxidation Reformer, *Int. J. Hydr. Energy* 2008, *33*, 7074–7083.

[25] Yuan, X., Liu, J., Zeng, G., Shi, J., Tong, J., Huang, G. Optimization of Conversion of Waste Rapeseed Oil With High FFA to Biodiesel Using Response Surface Methodology, *Renew. Energy* 2008, *33*, 1678–1684.

[26] Eide, I., Zahlsen, K. Chemical Fingerprinting of Biodiesel Using Electrospray Mass Spectrometry and Chemometrics: Characterization, Discrimination, Identification, and Quantification in Petrodiesel, *Energy Fuels* 2007, *21*, 3702–3708.

[27] Ortiz-Martinez, M., Winkler, R., García-Lara, S. Preventive and Therapeutic Potential of Peptidesfrom Cereals Against Cancer, *J. Proteom.* 2014, *111*, 165–183.

[28] Zhang, S., Wang, Z., Xu, S., Gao, X. Purification and Characterization of a Radical Scavenging Peptide from Rapeseed Protein Hydrolysates, *J. Am. Oil Chem. Soc.* 2009, *86*, 959–966.

[29] Koski, A., Pekkarinen, S., Hopia, A., Waehaelae, K., Heinonen, K. Processing of Rapeseed Oil: Effects on Sinapic Acid Derivative Content and Oxidative Stability, *Eur. Food Res. Technol.* 2003, *217*, 110–114.

[30] Yamamoto, K., Shibahara, A., Nakayama, T., Kajimoto, G. Determination of Double-bond Positions in Methylene-interrupted Dienoic Fatty Acids by GC–MS as Their Dimethyl Disulfide Adducts, *Chem. Phys. Lipids* 1991, *60*, 39–50.

[31] Holcapek, M., Jandera, P., Zderadicka, P., Hruba, L. Characterization of Triacylglycerol and Diacylglycerol Composition of Plant Oils Using High-performance Liquid Chromatography–atmospheric Pressure Chemical Ionization Mass Spectrometry, *J. Chromatogr. A* 2003, *1010*, 195–215.

[32] Kamal-Eldin, A., Marquez-Ruiz, G., Dobarganes, C., Appelqvist, L. Characterisation of Aldehydic Acids in Used and Unused Frying Oils, *J. Chromatogr. A* 1997, *776*, 245–254.

[33] Mildner-Szkudlarz, S., Jelen, H., Zawirska-Wojtasiak, R., Wcasowicz, E. Application of Headspace–solid Phase Microextraction and Multivariate Analysis for Plant Oils Differentiation, *Food Chem.* 2003, *83*, 515–522.

[34] Boukhchina, S., Sebai, K., Cherif, A., Kallel, H., Mayer, P. Identification of Glycerophospholipids in Rapeseed, Olive, Almond, and Sunflower Oils by LC–MS and LC–MS–MS, *Can. J. Chem.* 2004, *82*, 1210–1215.

[35] Lambelet, P., Grandgirard, A., Gregoire, S., Juaneda, P., Sebedio, J., Bertoli, C. Formation of Modified Fatty Acids and Oxyphytosterols during Refining of Low Erucic Acid Rapeseed Oil, *J. Agric. Food Chem.* 2003, *51*, 4284–4290.

[36] Kallio, H., Johansson, J., Oksmant, P. Composition and Development of Turnip Rapeseed (*Brassica campestris*) Oil Triacylglycerols at Different Stages of Maturation, *J. Agric. Food Chem.* 1991, *39*, 1752–1756.

[37] Wolff, R., Sebedio, J. Geometrical Isomers of Linolenic Acid in Low-Calorie Spreads Marketed in France, *J. Am. Oil Chem. Soc.* 1991, *68*, 719–725.

[38] Lin, C., Xie, X., Fan, N., Tu, Y., Chen., Y., Liao, W. Fast Analysis of Common Fatty Acids in Edible Vegetable Oils by Ultra-performance

Chromatography-mass Spectrometry, *Ch. J. Cromatogr.* 2015, *33*, 397–400.

[39] Durant, A., Dumont, M., Narine, S. In situ Silylation for the Multicomponent Analysis of Canola Oil by-products by Gas Chromatography–Mass Spectrometry, *Anal. Chim. Acta* 2006, *559*, 227–233.

[40] Li, J., Yue, Y., Hu, X., Zhong, H. Rapid Transmethylation and Stable Isotope Labeling for Comparative Analysis of Fatty Acids by Mass Spectrometry, *Anal. Chem.* 2009, *81*, 5080–5087.

[41] Beermann, C., Winterling, N., Green, A., Moebius, M., Schmitt, J., Boehm, G. Comparison of the Structures of Triacylglycerols from Native and Transgenic Medium-Chain Fatty Acid-Enriched Rape Seed Oil by Liquid Chromatography–Atmospheric Pressure Chemical Ionization Ion-Trap Mass Spectrometry (LC–APCI–ITMS) *Lipids* 2007, *42*, 383–394.

[42] Kalo, P., Ollilainen, V., Rocha, J., Malcata, F. Identification of Molecular species of Simple Lipids by Normal Phase Liquid Chromatography–Positive Electrospray Tandem Mass Spectrometry, and Application of Developed Methods in Comprehensive Analysis of Low Erucic Acid Rapeseed Oil Lipids, *Int. J. Mass Spectrom.* 2006, *254*, 106–121.

[43] Berdeaux, O., Gregoire, S., Fournier, C., Christie, W., Lambelet, P., Sebediod, J. Detection of Lactobacillic Acid in Low Erucic Rapeseed Oil–A Note of Caution when Quantifying Cyclic Fatty Acid Monomers in Vegetable Oils, *Chem. Phys. Lip.* 2010, *163*, 698–702.

[44] Lesknen, H., Suomela, J., Kallio, H. Quantification of Triacylglycerol Regioisomers by Ultra-high-performance Liquid Chromatography and Ammonia Negative Ion Atmospheric Pressure Chemical Ionization Tandem Mass Spectrometry, *Rapid Commun. Mass Spectrom.* 2010, *24*, 1–5.

[45] Mottram, H., Woodbury, S., Evershed, R. Identification of Triacylglycerol Positional Isomers Present in Vegetable Oils by High Performance Liquid Chromatography/Atmospheric Pressure Chemical Ionization Mass Spectrometry, *Rapid Commun. Mass Spectrom.* 1997, *11*, 1240–1252.

[46] Pitts, S., Thomson, C., Analysis and Classification of Common Vegetable Oils, *J. Forensic Sci.* 2003, *48*, 1293–1297.

[47] Svensson, J., Adlercreutz, P. Identification of Triacylglycerols in the Enzymatic Transesterification of Rapeseed and Butter Oil, *Eur. J. Lipid Sci. Technol.* 2008, *110*, 1007–1013.
[48] Boettcher, C., Roepenack-Lahaye, E., Schmidt, J., Ciemens, S., Scheel, D. Analysis of phenolic choline esters from seeds of *Arabidopsis thaliana* and *Brassica napus* by capillary Liquid Chromatography/ electrospray-tandem mass spectrometry, *J. Mass Spectrom.* 2009, *44*, 466–476.
[49] Christie W., Breckenridge G., Separation of *Cis* and *Trans* isomers of Unsaturated fatty Acids by High-Performance Liquid Chromatography in the Silver Ion Mode, *J. Chromatogr.* 1989, *469*, 261–269.
[50] Zeb, A., Murkovic, M. Analysis of Triacylglycerols in Refined Edible Oils by Isocratic HPLC–ESI–MS, *Eur. J. Lipid Sci. Technol.* 2010, *112,* 844–851.
[51] Richter, E., Spangenberg, J., Kreuzer, M., Leiber, F. Characterization of Rapeseed (*Brassica napus*) Oils by Bulk C, O, H, and Fatty Acid C Stable Isotope Analyses, *J. Agric. Food Chem.* 2010, *58*, 8048–8055.
[52] Vetter, W., Schroeder, M., Lehnert, M. Differentiation of Refined and Virgin Edible Oils by Means of the *Trans*- and *Cis*-Phytol Isomer Distribution, *J. Agric. Food Chem.* 2012, *60*, 6103–6107.
[53] Borisjuk, L., Neuberger, T., Schwender, J., Heinzel, N., Sunderhaus, S., Fuchs, J., Hay, J., Tschiersch, H., Braun, H., Denolf, P., Lambert, B., Jakob, P., Rolletscheka, H. Seed Architecture Shapes Embryo Metabolism in Oilseed Rape, *Plant Cell* 2013, *25*, 1625–1640.
[54] Schroeder, M., Vetter, W. Investigation of Unsaponifiable Matter of Plant Oils and Isolation of Eightphytosterols by Means of High-speed Counter-current Chromatography, *J. Chromatogr. A* 2012, *1237*, 96–105.
[55] Plutowska, B., Chmie, T., Dymerski, T., Wardencki, W. A Headspace Solid-phase Microextraction Method Development and Itsapplication in the Determination of Volatiles in Honeys by Gas Chromatography, *Food Chem.* 2011, *126*, 1288–1298.
[56] Kim, H., Salem, N. Liqid Chromatography Mass–Spectrometry of Lipids, *Prog. Lipid Res.* 1993, *32*, 221–245.
[57] Gutnikov, G. Fatty Acid Profiles of Lipid Samples, *J. Chromatogr. B* 1995, *671*, 71–89.
[58] Christie, W. Silver Ion and Chiral Chromatography in the Analysis of Triacylglycerols, *Prog. Lipid Res.* 1994, *33*, 9–18.

[59] Christie, W. Separation of Molecular Species of Triacylglycerols by High-Performance Liquid Chromatography with a Silver Ion Column, *J. Chromatogr.* 1988, *454*, 273–284.

[60] Schuyla, P., De Joodea, T., Vasconcellos, M., Duchateau, G. Silver-phase High-performance Liquid Chromatography–electrospray Mass Spectrometry of Triacylglycerols, *J. Chromatogr. A* 1998, *810*, 53–61.

[61] Spitzer, V. Screening Analysis of Unknown Seed Oils, *Fett/Lipid* 1999, *101*, 2–19.

[62] Sandra, P., Dermaux, A., Ferraz, V., Dittmann, M., Rozing, G. Analysis of Triglycerides by Capillary Electrochromatography, *J. Micro. Sep.* 1997, *9*, 409–419.

[63] Juaneda, P., Sebedio, J., Christie, W. Complete Separation of the Geometrical Isomers of Linoienic Acid by High Performance Liquid Chromatography with a Silver Ion Column, *J. High Res. Chromatogr.* 1994, *17*, 321–324.

[64] Byrdwell, W., Neff, W. Analysis of Hydroxy-Containing Seed Oils Using Atmospheric Pressure Chemical Ionization Mass Spectrometry, *J. Liq. Chromatogr. Rel. Technol.* 2011, *21*, 1485–1501.

[65] Duffin, K., Henion, J. Drug Testing and Toxicology, W. Drive, J. Shieh, Electrospray and Tandem Mass Spectrometric of Acylglycerol Mixtures That Are Dissolved in Solvents, *Anal. Chem.* 1991, *83*, 1781–1788.

[66] Laakso, P., Voutilainen, P. Analysis of Triacylglycerols by Silver-ion High-Performance Chemical Ionization Mass Spectrometry, *Lipids* 1996, *31*, 1311–1322.

[67] Waters, K., Ntambi, J. Spectrometric analysis of Triacylglycerols Using Negative–Ion Chemical Ionization with Ammonia, *Lipids* 1996, *31*, 33–42.

[68] Hsu, F., Turk, J. Structural Characterization of Triacylglycerols as Lithiated Adducts by Electrospray Ionization Mass Spectrometry Using Low–Energy Collisionally Activated Dissociation on a Triple Stage Quadrupole Instrument, *J. Am. Soc. Mass Spectrom.* 1999, *10*, 587–599.

[69] Ji, H., Voinov, V., Deinzer, M., Barofsky, D. Distinguishing between *Cis/Trans* Isomers of Monounsaturated Fatty Acids by FAB MS, *Anal. Chem.* 2007, *79*, 1519–1522.

[70] Wu, Z., Rodgers, R., Marshall, A. Characterization of Vegetable Oils: Detailed Compositional Fingerprints Derived from Electrospray Ionization Fourier Transform Ion Cyclotron Resonance Mass Spectrometry, *J. Agric. Food Chem.* 2004, *52*, 5322–5328.

[71] Cornett, D., Reyzer, M., Chaurand, M., Caprioli, R. MALDI Imaging Mass Spectrometry: Molecular Snapshots of Biochemical Systems, *Nat. Met.* 2007, *4*, 828–826.

[72] Sinha, T., Khatib-Shahidi, S., Yankeelov, T., Mapara, K., Ehtesham, M., Cornett, D., Dawant, D., Caprioli, R., Gore, J. Integrating Spatially Resolved Threedimensional MALDI–IMS with *in vivo* Magnetic Resonance Imaging, *Nat. Met.* 2008, *5*, 57–64.

[73] Seeley, E., Caprioli, R. MALDI Imaging Mass Spectrometry of Human Tissue: Method Challenges and Clinical Perspectives, *Trends Biotechnol.* 2011, *29*, 136–143.

[74] Norris, J., Caprioli, R. Analysis of Tissue Specimens by Matrix-assisted Laser Desorption/ionization Imaging Mass Spectrometry in Biological and Clinical Research, *Chem. Rev.* 2013, *113*, 2309–2342.

[75] Ivanova, B., Spiteller, M. Solid–state Maldi Mass Spectromery for Environmental Aanalysis, Lambert Academic Publishing, 2015, Saarbruecken, pp. 1–358.

[76] Cole, R. (Ed.) Electrospray and MALDI mass spectrometry: Findamentals, Instrumentation, Practicalities and Biomedical application, Wiley, 2010, 2nd Ed., Hoboken, New Jersey, pp. 1–847.

[77] Kelly, S., Parker, I., Sharman, M., Dennis, J., Goodall, I. Assessing the authenticity of single oils using fatty acid stable carbon ($^{13}C/^{12}C$), *Food Chem.* 1977, *59*, 181–186.

[78] Qvist, I., Von Sydow, E. Unconventional Proteins Aroma Precursors, *J. Agric. Food Chem.* 1976, *24*, 437–442.

[79] Rico, M., Bruix, M., Gonzalez, C., Monsalve, R., Rodriguez, R. ^{1}H NMR Assignment and Global Fold of Napin BnIb, a Representative 2S Albumin Seed Protein, *Biochem.* 1996, *35*, 15672–15682.

[80] Muren, E., Ek, B., Bjork, I., Rask, L. Structural Comparison of the Precursor and the Mature form of Napin, the 2s Storage Protein in *Brassica napus*, *Eur. J. Biochem.* 1996, *242*, 214–219.

[81] Barciszewski, J., Szymanski, M., Haertle, T. Minireview: Analysis of Rape Seed Napin Structure and Potential Roles of the Storage Protein, *J. Prot. Chem.* 2000, *19*, 249–254.

[82] Krzyzaniak A., Burova T., Haertle T., Barciszewski J., The Structure and Properties of Napin-seed Storage Protein from rape (*Brassica napus L.*), *Nahrung* 1998, *42,* 201–204.

[83] Schmidt, I., Renard, D., Rondeau, D., Richomme, P., Popineau, Y., Axelos, M. Detailed Physicochemical Characterization of the 2S Storage

Protein from Rape (*Brassica napus L.*), *J. Agric. Food Chem.* 2004, *52*, 5995–6001.

[84] Puumalainen, T., Puustinen, B., Poikonen, S., Turjanma, K., Palosuo, T., Vaali, K. Proteomic Identification of Allergenic Seed Proteins, Napin and Cruciferin, from Cold-pressed Rapeseed Oils, *Food Chem.* 2015, *175*, 381–385.

[85] Ishikawa, S., Maruyama, A., Yamamoto, Y., Hara, S. Extraction and Characterization of Glucosinolates and Isothiocyanates from Rape Seed Meal, *J. Oleo Sci.* 2014, *63*, 303–308.

[86] Lee, J., Bonnema, G., Zhang, N., Kwak, J., De Vos, R., Beekwilder, J. Evaluation of Glucosinolate Variation in a Collection of Turnip (*Brassica rapa*) Germplasm by the Analysis of Intact and DesulfoGlucosinolates, *J. Agric. Food Chem.* 2013, *61*, 3984–3993.

[87] Del Carpio, D., Basnet, R., Arends, D., Lin, K., De Vos, Muth, R., Kodde, J., Boutilier, K., Bucher, J., Wang, X. Ritsert Jansen5,Guusje Bonnema, Regulatory Network of Secondary Metabolism in Brassicarapa: Insight into the Glucosinolate Pathway, *Plos One* 2014, *9*, e107123 (12 pages).

[88] Song, L., Morrison, J., Botting, N., Thornalley, P. Analysis of Glucosinolates, Isothiocyanates, and Amine Degradationproducts in Vegetable Extracts and Blood Plasma by LC–MS/MS, *Anal. Biochem.* 2005, *347*, 234–243.

[89] Rochfort, R., Trenerry, V., Imsic, M., Panozzo, J., Jones, R. Class Targeted Metabolomics: ESI ion Trap Screening Methodsfor Glucosinolates Based on MSn Fragmentation, *Phytochem.* 2008, *69*, 1671–1679.

[90] Millan, S., Sampedro, M., Gallejones, P., Castellon, A., Ibargoitia, M., Goicolea, M., Barrio, R. Identification and Quantification of Glucosinolatesin Rapeseed Using Liquid Chromatography–ion Trapmass Spectrometry, *Anal. Bioanal. Chem.* 2009, *394*, 1661–1669.

[91] Guo, W., Zhou, H., Zhou, H., Chen, P., Guo, G. Glucosinolates in Rapeseeds by Liquid Chromatography–Electrospray Mass Spectrometry, *Anal. Lett.* 2005, *38*, 343–351.

[92] B. Auger, J. Pouvreau, K. Pouponneau, K. Yoneyama, G. Montiel, B. Le Bizec, K. Yoneyama, P. Delavault, R. Delourme, P. Simier, Germination Stimulants of *Phelipanche ramosa* in the Rhizosphere of *Brassica napus* Are Derived from the Glucosinolate Pathway, *Mol. Plant-Microbe Inter.* 2012, *25*, 993–1004.

[93] Gehrig, P., Krzyzaniaki, A., Barciszewski, J., Biemann, K. Mass Spectrometric Amino Acid Sequencing of a Mixture of Seed Storage Proteins (napin) from *Brassica napus*, Products of a Multigene Family, *Proc. Natl. Acad. Sci. USA* 1996, *93*, 3647–3652.

[94] Puumalainen, P., Poikonen, S., Kotovuori, A., Vaali, K., Kalkkinen, N., Reunal, T., Turjanma, K., Palosuo, T. Napins, 2S albumins, are Major Allergens in Oilseed Rape and Turnip Rape, *J. Allergy. Clin. Immunol.* 2006, *117*, 426–32.

[95] Sheoran, I., Pedersen, E., Ross, A., Sawhney, V. Dynamics of protein expression during pollen germination in canola (*Brassica napus*), *Planta* 2009, *230*, 779–793.

[96] Wang, C., Li, B., Wang, B., Xie, N. Degradation and Antioxidant Activities of Peptides and Zinc–Peptidecomplexes During *in vitro* Gastrointestinal Digestion, *Food Chem.* 2015, *173*, 733–740.

[97] Srivastava, S., Fristensky, B., Kav, N. Constitutive Expression of a PR10 Protein Enhances the Germination of *Brassica napus* under Saline Conditions, *Plant Cell Physiol.* 2004, *45*, 1320–1324.

[98] Wilen, W., Mandel, R., Pharis, R., Holbrook, L., Moloney, M. Effects of Abscisic Acid and High Osmoticum on Storage rotein Gene Expression in Microspore Embryos of *Brassica napus*, *Plant Physiol.* 1990, *94*, 875–881.

[99] Gan, L., Zhang, C., Wang, X., Wang, H., Long, Y., Yin, Y., Li, D., Tian, J., Li, Z., Lin, Z., Yu, L., Li, M. Proteomic and Comparative Genomic Analysis of Two *Brassica napus* Lines Differing in Oil Content, *J. Proteome Res.* 2013, *12*, 4965–4978.

[100] Houston, N., Hajduch, M., Thelen, J. Quantitative Proteomics of Seed Filling in Castor:Comparison with Soybean and Rapeseed Reveals Differences between Photosynthetic and Nonphotosynthetic Seed Metabolism, *Plant Physiol.* 2009, *151*, 857–868.

[101] He, R., Malomo, S., Alashi, A., Girgih, A., Ju, X., Aluko, R. Purification and Hypotensive Activity of Rapeseed Proteinderived Renin and Angiotensin Converting Enzyme Inhibitory Peptides, *J. Func. Foods* 2013, 781–789.

[102] Tessier, B., Schweizer, M., Fournier, F., Framboisier, X., Chevalot, I., Vanderesse, R., Harscoat, C., Marc, I. Prediction of the Amino acid Composition of Small Peptides Contained in a Plant Protein Hydrolysate by LC–MS and CE–MS, *Food Res. Int.* 2005, *38*, 577–584.

[103] Albertin, W., Balliau, T., Brabant, P., Chevre, A., Eber, F., Malosse, C., Thiellement, H. Numerous and Rapid Nonstochastic Modifications of

Gene Products inNewly Synthesized Brassica napus Allotetraploids, *Gen.* 2006, *173*, 1101–1113.

[104] Schwender, J., Shachar-Hill, Y., Ohlrogge, J. Mitochondrial Metabolism in Developing Embryos ofBrassica napus, *J. Biol. Chem.* 2006, *281*, 34040–34047.

[105] Agrawal, G., Thelen, J. Large Scale Identification and Quantitative Profiling of Phosphoproteins Expressed duringSeed Filling in Oilseed Rape, *Mol. Cell. Proteom.* 2006, *5*, 2044–2059.

[106] Zhu, M., Dai, S., McClung, S., Yan, X., Chen, S. Functional Differentiation of *Brassica napus* Guard Cells and Mesophyll Cells Revealed by Comparative Proteomics, *Mol. Cell. Proteom.* 2009, *8*, 752–766.

[107] Cheng, Z., Duan, J., Hao, Y., McConkey, B., Glick, B. Identification of Bacterial Proteins Mediatingthe Interactions Between Pseudomonas putida UW4 and *Brassica napus* (Canola), *Mol. Plant-Microbe Inter.* 2009, *22*, 686–694.

[108] Agrawal, G., Hajduch, M., Graham, K., Thelen, J. In-Depth Investigation of the Soybean Seed-Filling Proteome and Comparison with a Parallel Studyof Rapeseed, *Plant Physiol.* 2008, *148*, 504–518.

[109] Devouge, V., Rogniaux, H., Nesi, N., Tessier, D., Gueguen, J., Larre, C. Differential Proteomic Analysis of Four Near-Isogenic *Brassica napus* Varieties Bred for their Erucic Acid and Glucosinolate Contents, *J. Proteom. Res.* 2007, *6*, 1342–1353.

[110] Aran, M., Caporaletti, C., Senn, A., Tellez de Inon, M., Girotti, M., Llera, A., Wolosiuk, R. ATP-dependent Modulation and Autophosphorylation of rapeseed 2-Cys peroxiredoxin, *FEBS J.* 2008, *275*, 1450–1463.

[111] Thelen, J. Proteomics Tools and Resources for Investigating Protein Allergens in Oilseeds, *Regul. Toxicol. Pharmacol.* 2009, *54*, S41–S45.

[112] Guo, X., Tian, S., Small, D. Generation of Meat-like Flavourings From Enzymatic Hydrolysates of Proteins from *Brassica sp. Food Chem.* 2010, *119*, 167–172.

[113] Zhu, M., Simons, B., Zhu, N., Oppenheimer, D., Chen, S. Analysis of Abscisic Acid Responsive Proteins in *Brassica napus* guard cells by Multiplexed Isobaric Tagging, *J. Proteom.* 2010, *73*, 790–805.

[114] Harscoat-Schiavo, C., Nioi, C., Ronat-Heit, E., Paris, C., Vanderesse, R., Fournier, F., Marc, I. Hydrophilic Properties as a New Contribution for Computer-aided Identification of Short Peptides in Complex Mixtures, *Anal. Bioanal. Chem.* 2012, *403*, 1939–1949.

[115] Yao, Y., Sun, H., Xu, F., Zhang, X., Liu, S. Comparative Proteome Analysis of Metabolic Changes by Low Phosphorus Stress in two *Brassica napus* Genotypes, *Planta* 2011, *233*, 523–537.

[116] Kong, F., Mao, S., Du, K., Wu, M., Zhou, X., Chu, C., Wang, Y. Comparative Proteomics Analysis of OsNAS1 transgenic *Brassica napus* under salt stress, *Chin. Sci. Bull.* 2011, *56*, 2343–2350.

[117] Tan, X., Wang, Q., Tian, B., Zhang, H., Lu, D., Zhou, J. A *Brassica napus* Lipase Locates at the Membrane Contact Sites Involved in Chloroplast Development, *PloS ONE* 2011, *6*, e26831 (10 pages).

[118] Clauss, K., Von Roepenack-Lahaye, E., Boettcher, C., Roth, M., Welti, R., Erban, A., Kopka, J., Scheel, D., Milkowski, C., Strack, D. Overexpression of Sinapine Esterase BnSCE3 in Oilseed Rape Seeds Triggers Global Changes in Seed Metabolism, *Plant Physiol.* 2011, *155*, 1127–1145.

[119] Demartini, D., Jain, R., Agrawal, G., Thelen, J. Proteomic Comparison of Plastids from Developing Embryos and Leaves of Brassica napus, *J. Proteome Res.* 2011, *10*, 2226–2237.

[120] Zhu, M., Dai, S., Zhu, N., Booy, A., Simons, B., Yi, S., Chen, S. Methyl Jasmonate Responsive Proteins in Brassica napus Guard Cells Revealed by iTRAQ-Based Quantitative Proteomics, *J. Proteome Res.* 2012, *11*, 3728–3742.

[121] Gan, L., Li, D., Zang, X., Fu, C., Yu, L., Li, M. Construction of Two-Dimensional Polyacrylamide Gel Electrophoresis System for Proteins in *Brassica napus*, *Acta Agron. Sinica* 2010, *36*, 612–619.

[122] Velasco, P., Francisco, M., Moreno, D., Ferreres, F., García-Viguera, C., Cartea, M. Phytochemical Fingerprinting of Vegetable *Brassica oleracea* and *Brassica napus* by Simultaneous Identification of Glucosinolates and Phenolics, *Phytochem. Anal.* 2011, *22*, 144–152.

[123] Fan, S., Wang, X., Li, P., Zhang, Q., Zhang, W. Simultaneous Determination of 13 Phytohormones in Oilseed rape Tissues by Liquid Chromatography-electrospray Tandem Mass Spectrometry and the Evaluation of The Matrix Effect, *J. Sep. Sci.* 2011, *34*, 640-650.

[124] Li, Y., Wei, F., Dong, X., Peng, J., Liu, S., Chen, H. Simultaneous Analysis of Multiple Endogenous Plant Hormones in Leaf Tissue of Oilseed rape by Solid-phase extraction Coupled with High-performance Liquid Chromatography-electrospray Ionisation Tandem Mass Spectrometry, *Phytochem. Anal.* 2011, *22*, 442–449.

[125] Jiang, J., Wang, J., Li, A., Zhang, Y., Sokolov, V., Wang, Y. Proteomic differences in seed filling between yellow-seeded progeny of *Brassica*

napus-Sinapis alba (*Brassicaceae*) and black-seeded parent B. napus, *Russ. J. Gen.* 2012, *48*, 396–403.

[126] Jiang, J., Wang, J., Li, A., Zhang, Y., Sokolov, V., Wang, Y. Proteomic differences in seed filling between yellow-seeded progeny of *Brassica napus-Sinapis alba* (*Brassicaceae*) and black-seeded parent *B. napus*. *Genetika* 2012, *48*, 480–487.

[127] Higgins, J., Magusin, A., Trick, M., Fraser, F., Bancroft, I. Use of mRNA-seq to Discriminate Contributions to the Transcriptome from the Constituent Genomes of the Polyploid crop species *Brassica napus*, *BMC Genom.* 2012, *13*, Article number 247.

[128] Giavalisco, P., Kapitza, K., Kolasa, A., Buhtz, A., Kehr, J. Towards the Proteome of *Brassica napus* phloem sap, *Proteom.* 2006, *6*, 896–909.

[129] Katavic, V., Agrawal, G., Hajduch, M., Harris, S., Thelen, J. Protein and Lipid Composition Analysis of oil Bodies from two *Brassica napus* cultivars, *Proteom.* 2006, *6*, 4586–4598.

[130] Jolivet, P., Tailliart, K., Boulard, C., Nesi, N., Chardot, T. Purification and protein composition of oil bodies from *Brassica napus* seeds, *Oleag. Corps Gras Lipides* 2006, *13*, 426–430.

[131] Hajduch, M., Agrawal, G., Thelen, J. Proteomics of Seed Development in Oilseed Crops, *Plant Proteom.* 2007, 137–154.

[132] Leskinen, H., Suomela, J., Kallio, H. Quantification of Triacylglycerol Regioisomers in Oils and Fat using Different Mass Spectrometric and Liquid Chromatographic Methods, *Rapid Commun. Mass Spectrom.* 2007, *21*, 2361–2373.

[133] Selvam, A., Wong, J. Phytochelatin Systhesis and Cadmium Uptake of *Brassica napus*, *Environm. Technol.* 2008, *29,* 765–773.

[134] Liang, Y., Strelkov, S., Kav, N. Oxalic acid-mediated Stress Responses in *Brassica napus* L., *Proteom.* 2009, *9*, 3156–3173.

[135] Jolivet, P., Boulard, C., Bellamy, A., Larre, C., Barre, M., Rogniaux, H., D'Andrea, S., Chardot, T., Nesi, N. Protein Composition of Oil Bodies from Mature *Brassica napus* seeds, *Proteom.* 2009, *9*, 3268–3284.

[136] Thelen, J. Proteomics Tools and Resources for Investigating Protein Allergens in Oilseeds, *Regul. Toxicol. Pharmacol.* 2009, *54*, S41–S45.

[137] Chena, S., Ding, G., Wang, Z., Cai, H., Xu, F. Proteomic and Comparative Genomic Analysis Reveals Adaptability of *Brassica napus* tophosphorus-deficient Stress, *J. Proteom.* 2015, *117*, 106–119.

[138] Tarayre, C., Brognaux, A., Brasseur, C., Bauwens, J., Millet, C., Matteotti, C., Destain, J., Vandenbol, M. , Portetelle, D., De Pauw, E., Haubruge, E., Francis, F., Thonart, P. Isolation and Cultivation of a

Xylanolytic Bacillus subtilisExtracted from the Gut of the TermiteReticulitermes santonensis, *Appl. Biochem. Biotechnol.* 2013, *171*, 225–245.

[139] Berot, S., Compoint, J., Larre, C., Malabat, C., Gueguen, J. Large scale purification of rapeseed proteins (*Brassica napus L.*), *J. Chromatogr. B* 2005, *818*, 35–42.

[140] Weselake, R., Madhavji, M., Szarka, S., Patterson, N., Wiehler, W., Nykiforuk, C., Burton, T., Boora, P., Mosimann, S., Foroud, N., Thibault, B., Moloney, M., Laroche, A., Furukawa-Stoffer, T. Acyl-Co A-binding and Self-associating Properties of a Recombinant 13.3 kDa N-terminal Fragment of Diacylglycerol Acyltransferase-1 from Oilseed rape, *BMC Biochem.* 2006, *7*, 24 (13 pages).

[141] Albertin, W., Alix, K., Balliau, T., Brabant, P., Davanture, D., Malosse, C., Valot, B., Thiellement, H. Differential Regulation of Gene Products in Newly Synthesized *Brassica napus* Allotetraploids is not Related to Protein Function Norsubcellular Localization, *BMC Genom.* 2007, *8*, 56 (15 pages).

[142] Kehr, J., Buhtz, A., Giavalisco, P. Analysis of Xylem sap Proteins from *Brassica napus*, *BMC Plant Biol.* 2005, *5*, 11 (13 pages).

[143] Krygier, K., Sosulski, F., Hogge, L. Free, Esterified, and Insoluble-Bound Phenolic Acids. 2. Composition of Phenolic Acids in Rapeseed Flour and Hulls, *J. Agric. Food Chem.* 1982, *30*, 334–336.

[144] Itoh, T., Tamura, T., Matsumoto, T. Sterol Composition of 19 Vegetable Oils, Lipids 1973, 8, 122–125.

[145] Blomstrand, R., Svensson, L., Herslof, B. Mass Fragmentographic Determination of Docosenoic Acid in Rapeseed Oils, *Lipids* 1977, *13*, 283–288.

[146] Blomstrand, R., Svensson, L. Studies on Phospholipids with Particular Reference to Cardiolipin of Rat Heart after Feeding Rapeseed Oil, *Lipids* 1974, *9*, 771–780.

[147] Latxague, L., Gardrat, C., Coustille, J., Viaud, M., Rollin, P. Identification of Enzymatic Degradation Products from Synthesized Glucobrassicin by Gas Chromatography-mass Spectrometry, *J. Chromatogr.* 1991, *586*, 166–1 70

[148] Truscott, R., Burke, D., Minchinton, I. The Characterization of as Novel Hydroxiindole Glucosinolate, Biochem. Biophys. *Res. Commun.* 1982, *107*, 1258–1264.

[149] Siddiqui, I., Wood, P., Khanzada, G. A di-D-galactosyl-myo-inositol from Rapeseed (*Brassica campestris*) Meal, *Carbohyd. Res.* 1973, *29*, 255–258.
[150] Siddiqui, I., Wood, P. Structural Investigation of Water Soluble Rapeseed (*Brasszca campestrrs*) Polysaccharides, Part II and Acidic Arabinogalactan, Carbohyd. Res. 1972, 24, l–9.
[151] Aspinall, G., Jung, K. Rapeseed Hull Pectin, *Carbohyd. Res.* 1974, *38*, 247–255.
[152] Aspinall, G., Krishnamurthy, T., Rosell, K. A Fucogalactoxyloglycan from Rapeseed Hull, *Carbohyd. Res.* 1977, *55*, 11–19.
[153] York, W., Van Halbeek, H., Darvill, A., Albersheimt, P. Structural Analysis of Xyloglucan Oligosaccharides by ^{1}H n.m.r. Spectroscopy and Fast-atom-bombardment Mass Spectrometry, *Carbohyd. Res.* 1990, *200*, 9–31.
[154] Selvaraj, G., Jain, R., Olson, D., Hirji, R., Jana, S., Hogge, L. Glycinebetaine in Oilseed Rape and Flax leaves: Detection by Liquid Chromatography Flow Secondary Ion Mass Spectrometry, *Phytochem.* 1995, *38*, 1143–1146.
[155] Zhou, Z., Squires, T., Ambrose, S., Abrams, S., Ross, A., Cutler, A. Rapid Extraction of Abscisic Acid and its Metabolites for Liquid Chromatography-tandem Mass Spectrometry, *J. Chromatogr. A* 2003, *1010*, 75–85.
[156] Ross, A., Ambrose, S., Cutler, A., Feurtado, F., Kermode, A., Nelson, K., Zhou, R., Abramsa, S. Determination of Endogenous and Supplied Deuterated Abscisic Acid in Plant Tissues by High-performance Liquid Chromatography-electrospray Ionization Tandem Mass Spectrometry with Multiple Reaction Monitoring, *Anal. Biochem.* 2004, *329*, 324–333.
[157] Tolra, R., Alonso, R., Poschenrieder, C., Barcelo, D., Barcelo, J. Determination of Glucosinolates in Rapeseed and *Thlaspi caerulescens* Plants by Liquid Chromatography–Atmospheric Pressure Chemical Ionization Mass Spectrometry, *J. Chromatogr. A* 2000, *889*, 75–81.
[158] Lampi, A., Juntunen, L., Toivo, T., Piironen, V. Determination of Thermo-oxidation Products of Plant Sterols, *J. Chromatogr. B*, 2002, *777*, 83–92.
[159] Calaf, R., Pena, J., Paytubi, S., Blount, B., De la Paz, M., Gelpi, E., Abian, J. Determination of Aniline Derivatives in Oils Related to the Toxic Oil Syndrome by Atmospheric Pressure Ionization-Tandem Mass Spectrometry, *Anal. Chem.* 2001, *73*, 3828–3837.

[160] Lindh L., Dahlen J., Hydrogen Sulfide and Acetaldehyde Discharge from a Rapeseed Extraction Plant, *J. Am. Oil Chem. Soc.* 1989, *66*, 972–973.

[161] Zhu X., Wang K., Zhu J., Koga M., Analysis of Cooking Oil Fumes by Ultraviolet Spectrometry and Gas Chromatography-Mass Spectrometry, *J. Agric. Food Chem.* 2001, *49*, 4790–4794.

[162] Shields P., Xu G., Blot W., Fraumeni J. Jr., Trivers G., Pellizzari E., Qu Y., Gao Y., Harris C., Mutagens From Heated Chinese and U.S. Cooking Oils, *J. Nat. Cancer Inst.* 1995, *87*, 836–841.

[163] Jelen H., Muchowska M., Zawirska–Wojtasiak R., Wauasowicz E., Headspace Solid-Phase Microextraction Use for the Characterization of Volatile Compounds in Vegetable Oils of Different Sensory Quality, *J. Agric. Food Chem.* 2000, *48*, 2360–2367.

[164] Grigorakis K., Fountoulaki E., Giogios I., Alexis M., Volatile Compounds and Organoleptic Qualities of Gilthead Sea Bream (*Sparus aurata*) Fed Commercial Diets Containing Different Lipid Sources, *Aquaculture* 2009, *290*, 116–121.

[165] Wolfram, K., Schmidt, J., Wray, V., Milkowski, C., Schliemann, W., Strack, D. Profiling of Phenylpropanoids in Transgenic Low-sinapine Oilseed Rape (*Brassica napus*), *Phytochem.* 2010, *71*, 1076–1084.

[166] Harbaum-Piayda, B., Oehlke, K., Soennichsen, F., Zacchi, P., Eggers, R., Schwarz, K. New Polyphenolic Compounds in Commercial Deodistillate and Rapeseed Oils, *Food Chem.* 2010, *123*, 607–615.

[167] Auger, B., Marnet, N., Gautier, V., Maia-Groundard, A., Leprince, F., Renard, M., Guyot, S., Nesi, N., Routaboul, J. A Detailed Survey of Seed Coat Flavonoids in Developing Seeds of *Brassica napus* L., *J. Agric. Food Chem.* 2010, *58*, 6246–6256.

[168] Tarvainen, M., Phuphusit, A., Suomela, J., Kuksis, A., Kallio, H. Effects of Antioxidants on Rapeseed Oil Oxidation in an Artificial Digestion Model Analyzed by UHPLC–ESI–MS, *J. Agric. Food Chem.* 2012, *60*, 3564–3579.

[169] Frolov, A., Henning, A., Boettcher, C., Tissier, A., Strack, D. An UPLC-MS/MS Method for the Simultaneous Identification and Quantitation of Cell Wall Phenolics in *Brassica napus* Seeds, *J. Agric. Food Chem.* 2013, *61*, 1219–1227.

[170] Watanabe, A., Okazaki, K., Watanabe, T., Osaki, M., Shinano, T. Metabolite Profiling of Mizuna (*Brassica rapa L.* var. Nipponsinica) To Evaluate the Effects of Organic Matter Amendments, *J. Agric. Food Chem.* 2013, *61*, 1009–1016.

[171] Claussen, F., Taylor, M., Breeze, M., Liu, K. Measurement of Vitamin K1 in Commercial Canola Cultivars fromGrowing Locations in North and South America Using High-Performance Liquid Chromatography–Tandem Mass Spectrometry, *J. Agric. Food Chem.* 2015, *63*, 1076–1081.

[172] Shao, Y., Jiang, J., Ran, L., Lu, C., Wei, C., Wang, Y. Analysis of Flavonoids and Hydroxycinnamic Acid Derivatives in Rapeseeds (*Brassica napus L.* var. napus) by HPLC–PDA–ESI(–)–MSn/HRMS, *J. Agric. Food Chem.* 2014, *62*, 2935–2945.

[173] Fang, J., Reichelt, M., Kai, M., Schneider, B. Metabolic Profiling of Lignans and Other Secondary Metabolitesfrom Rapeseed (*Brassica napus L.*), *J. Agric. Food Chem.* 2012, *60*, 10523–10529.

[174] Qu, C., Fu, F., Lu, K., Zhang, K., Wang, R., Xu, X., Wang, M., Lu, J., Wan, H., Zhanglin, T., Li, J. Differential Accumulation of phenolic compounds andexpression of related genes in black- and yellow-seeded *Brassica napus*, *J. Exper. Bot.* 2013, *64*, 2885–2898.

[175] Fang, J., Reichelt, M., Hidalgo, W., Agnolet, S., Schneider, S. Tissue-Specific Distribution of Secondary Metabolites in Rapeseed (*Brassica napus* L.), PLoS ONE 2012, *7*, e48006 (9 pages).

[176] Zhou, Q., Yang, M., Huang, F., Zheng, C., Deng, Q. Effect of Pretreatment with Dehulling andMicrowaving on the Flavor Characteristicsof Cold-Pressed Rapeseed Oil by GC–MS–PCA and Electronic Nose Discrimination, *J. Food Sci.* 2013, *78*, C961–C970.

[177] Baumert, C., Milkowski, J., Schmidt, M., Nimtz, V., Wray, D., Strack, D. Formation of a complex pattern of sinapate esters inBrassica napus seeds, catalyzed by enzymes of a serinecarboxypeptidase-like acyltransferase family?, *Phytochem.* 2005, *66*, 1334–1345.

[178] Meyer L., Gao J., Xu D., Thelen J., Phosphoproteomic Analysis of Seed Maturation in Arabidopsis, Rapeseed, and Soybean, *Plant Physiol.* 2012, *159*, 517–528.

[179] Yakovlev, V., Zavarukhin, S., Kuzavov, V., Stebnovskii, S., Malykh, N., Maltsev, L., Parmon, V. A Study of Chemical Transformations of Organic Compounds Under the Action of Cavitation, *Russ. J. Phys. Chem. B* 2010, *4*, 227–234.

[180] Hong, E., Lim, C., Son, H., Choi, J., Soo, N., Rancidity Analysis of Rapeseed Oil Under Different Storage Conditions Using Mass Spectrometry-based Electronic Nose, *Kor. J. Food Sci. Technol.* 2010, *42*, 699–704.

[181] Zhao, P., Wang, L., Chen, L., Pan, C., Residue dynamics of clopyralid and picloram in rape plant rapeseed and field soil, *Bull. Environm. Contam. Toxicol.* 201, *86*, 78–82.

[182] Chae, L., Hong, E., Son, H., Kim, J., Noh, B. Effects of Roasting Condition and Storage Time on Changes in Volatile Compounds in Rapeseed Oils, *Kor. J. Food Sci. Technol.* 2011, *43*, 291–302.

[183] Uriarte, P., Goicoechea, E., Guillen, M. Volatile Components of Several Virgin and Refined Oils Differing in Their Botanical Origin, *J. Sci. Food and Agric.* 2011, *91*, 1871–1884.

[184] Plagemann, I., Zelena, K., Krings, U., Berger, R. Volatile Flavours in Raw Egg Yolk of Hens Fed on Different Diets, *J. Sci. Food and Agric.* 2011, *91*, 2061–2065.

[185] Gasparoli, A., Gaboardi, M. Analytical Evaluation of Sterols and Stanols in Hydrogenated Lipids, *Riv. Ital. Sost. Gras.* 2011, *88*, 219–229.

[186] Wang, X., Li, P., Zhang, W., Zhang, Q., Ma, F., Yu, L., Wang, L. Screening for Pesticide Residues in Oil Seeds Using Solid-phase Dispersion Extraction and Comprehensive Two-Dimensional Gas Chromatography Time-of-flight Mass Spectrometry, *J. Sep. Sci.* 2012, *35*, 1634–1643.

[187] Ma, F., Li, P., Matthäus, B., Zhang, W., Zhang, Q. Optimization of Ultrasonic-assisted Extraction of 3-monochloropropane-1,2-diol (MCPD) and Analysis of its Esters From Edible Oils by Gas Chromatography-mass Spectrometry, *J. Sep. Sci.* 2012, *35*, 2241–2248.

[188] Brumley, W., Sheppard, A., Rudolf, T., Shen, C., Yasaei, P., Sphon, J. Mass Spectrometry and Identification of Sterols in Vegetable Oils as Butyryl Esters and Relative Quantitation by Gas Chromatography with Flame Ionization Detection, *J. Assoc. Offic. Anal. Chem.* 1985, *68*, 701–709.

[189] Butcher, R., Macfarlane-Smith, W., Robertson, G., Griffiths, D. The Identification of Potential Aeroallergen/irritant(s) from Oilseed Rape (*Brassica napus* spp. oleifera): Volatile Organic Compounds Emitted During Flowering Progression, *Clin. Exper. Aller.* 1994, *24*, 1105–1114.

[190] Billaud, F., Dominguez, V., Broutin, P., Busson, C. Production of Hydrocarbons by Pyrolysis of Methyl Esters From Rapeseed Oil, *J. Am. Oil Chem. Soc.* 1995, *72*, 1149–1154.

[191] Dutta, P. Studies on Phytosterol Oxides. II: Content in Some Vegetable Oils and in French Fries Prepared in These Oils, *J. Am. Oil Chem. Soc.* 1997, *74*, 659–666.

[192] Juaneda, P., Cordier, O., Gregoire, S., Sebedio, J. Conjugated Linoleic Acid (CLA) Isomers in Heat-treated Vegetable Oils, *Oleag. Corps Gras Lipides* 2001, *8*, 94–97.
[193] Doleschall, F., Kemeny, Z., Recseg, K., Kovari, K. A new Analytical Method to Monitor Lipid Peroxidation During Bleaching, *Eur. J. Lipid Sci. Technol.* 2002, *104*, 14–18.
[194] Bellevik, S., Zhang, J., Meijer, J. *Brassica napus* Soluble Epoxide Hydrolase (BNSEH1): Cloning and Characterization of the Recombinant Enzyme Expressed in Pichia Pastoris, *Eur. J. Biochem.* 2002, *269*, 5295–5302.
[195] Hertenberger, G., Wanek, W. Evaluation of Methods to Measure Differential ^{15}N Labeling of Soil and Root N Pools for Studies of Root Exudation, *Rapid Commun. Mass Spectrom.* 2004, *18*, 2415–2425.
[196] Laurent, F., Debrauwer, L., Pascal-Lorber, S. Metabolism of [^{14}C]-2,4-dichlorophenol in edible plants, *Pest Manag. Sci.* 2006, *62*, 558–564.
[197] Reig, N., Calaf, R., Messeguer, A., Morato, A., Escabros, J., Gelpí, E., Abian, J. LC-MS Ion Maps for the Characterization of Aniline Derivatives of Fatty Acids and Triglycerides in Laboratory-denatured Rapeseed Oil, *J. Mass Spectrom.* 2007, *42*, 527–541.
[198] Crush, J., Briggs, L., Sprosen, J., Nichols, S. Effect of Irrigation with Lake Water Containing Microcystins on Microcystin Content and Growth of Ryegrass, Clover, Rape, and Lettuce, *Environm. Toxicol.* 2008, *23*, 246–252.
[199] Wagstaffe, P., Boenke, A., Schnug, E., Lindsey, A. Certification of the Sulphur Content of Three Rapeseed Reference Materials, *Fresenius J. Anal. Chem.* 1992, *344*, 1–7.
[200] Urbanova, T., Tarkowska, D., Novak, O., Hedden, P., Strnad, M. Analysis of Gibberellins as Free Acids by Ultra Performance Liquid Chromatography – Tandem Mass Spectrometry, *Talanta* 2013, *112*, 85–94.
[201] Cai, B., Zhu, J., Shi, Z., Yuan, B., Feng, Y. A Simple Sample Preparation Approach Based on Hydrophilicsolid-phase Extraction Coupled with Liquid Chromatography–tandemmass Spectrometry for Determination of Endogenous Cytokinins, *J. Chromatogr. B* 2013, *942–943*, 31–36.
[202] Liu, Q., Wu, L., Pu, H., Li, C., Hu, Q. Profile and distribution of soluble and insoluble phenolics in Chinese rapeseed (*Brassica napus*), *Food Chem.* 2012, *135*, 616–622.

[203] Vaclavik, L., Belkova, B., Reblova, Z., Riddellova, K., Hajslova, J. Rapid Monitoring of Heat-accelerated Reactions in Vegetable Oils Using Direct Analysis in Real Time Ionization Coupled with High Resolution Mass Spectrometry, *Food Chem.* 2013, *138*, 2312–2320.

[204] Aladedunye, F., Matthaeus, B. Phenolic Extracts from *Sorbus aucuparia* (L.) and *Malus baccata* (L.) Berries: Antioxidant Activity and Performance in Rapeseed Oilduring Frying and Storage, *Food Chem.* 2014, *159*, 273–281.

[205] Tarvainen, M., Nuora, A., Quirin, K., Kallio, H., Yang, B. Effects of CO_2 Plant Extracts on Triacylglycerol Oxidation in Atlanticsalmon During Cooking and Storage, *Food Chem.* 2015, *173*, 1011–1021.

[206] Soupas, L., Huikko, L., Lampi, A., Piironen, V. Pan-frying May Induce Phytosterol Oxidation, *Food Chem.* 2007, *101*, 286–297.

[207] Spielmeyer, A., Wagner, A., Jahreis, G. Influence of Thermal Treatment of Rapeseed on the Canolol Content, *Food Chem.* 2009, *112*, 944–948.

[208] Lanier, C., Heutte, N., Richard, E., Boucard, V., Lebailly, P., Garon, D. Mycoflora and Mycotoxin Production in Oilseed Cakes during Farm Storage, *J. Agric. Food Chem.* 2009, *57*, 1640–1645.

[209] Hori, K., Matsubara, A., Uchikata, T., Tsumura, K., Fukusaki, E., Bamba, T. High-throughput and sensitive analysis of 3-monochloropropane-1,2-diol fatty acid esters in edible oils by supercritical fluid chromatography/tandem mass spectrometry, *J. Chromatogr. A* 2012, *1250*, 99–104.

[210] Shrestha, K., Stevens, C., De Meulenaer, B. Isolation and Identification of a Potent Radical Scavenger (Canolol) from Roasted High Erucic Mustard Seed Oil from Nepal and Its Formation during Roasting, *J. Agric. Food Chem.* 2012, *60*, 7506–7512.

[211] Ucar, S., Ozkan, A. Characterization of products from the pyrolysis of rapeseed oil cake, *Bioresource Technol.* 2008, *99*, 8771–8776.

[212] Chabanon, I., Chevalot, X., Framboisier, S., Chenu, I., Hydrolysis of Rapeseed Protein Isolates: Kinetics, Characterization and Functional Properties of Hydrolysates, *Proc. Biochem.* 2007, *42*, 1419–1428.

[213] Qian, M., Zhang, H., Wu, L., Jin, N., Wang, J., Jiang, E. Simultaneous Determination of *Zearalenone* and its Derivatives in Ediblevegetable Oil by Gel Permeation Chromatography and Gas Chromatography–triple quadrupole mass spectrometry, *Food Chem.* 2015, *166*, 23–28.

[214] Drabova, L., Tomaniova, M., Kalachova, K., Kocourek, V., Hajslova, J., Pulkrabova, J. Application of Solid Phase Extraction and Two-dimensional Gas Chromatography Coupled with Time-of-flight Mass

Spectrometry for Fast Analysis of Polycyclic Aromatic Hydrocarbons in Vegetable Oils, *Food Contr.* 2013, *33*, 489–497.

[215] Yamazaki, K., Ogiso, M., Isagawa, S., Urushiyama, T., Ukena, T., Kibune, N. A new, Direct Analytical Method Using LC-MS/MS for Fatty Acid Esters of 3-chloro-1,2-propanediol (3-MCPD esters) in Edible Oils, *Food Add. Contam. A* 2013, *30*, 52–68.

[216] Pedras, M., Zheng, Q., Gadagi, R., Rimmer, S. Phytoalexins and Polar Metabolites from the Oilseeds Canola and Rapeseed: Differential Metabolic Responses to the *Biotroph Albugo Candida* and to Abiotic Stress, *Phytochem.* 2008, *69*, 894–910.

[217] Yamamoto, K., Kinoshita, A., Shibahara, A. Ricinoleic Acid in Common Vegetable Oils and Oil Seeds, *Lip.* 2008, *43*, 457–460.

[218] Zelinkova, Z., Svejkovska, B., Velisek, J., Dolezal, M. Fatty Acid Esters of 3-Chloropropane-1,2-diol in Edible Oils, *Food Add. Contam.* 2006, *23*, 1290–1298l.

[219] Pedras, M., Minic, Z., Jha, M. Brassinin Oxidase, a Fungal Detoxifying Enzyme to Overcome a Plant Defense–purification, Characterizationand Inhibition, *FEBS J.* 2008, *275*, 3691–3705.

[220] Tarkowska, D., Filek, M., Biesaga-Koscielniak, J., Marcinskac, I., Machackova, I., Krekule, J., Strnad, M. Cytokinins in Shoot Apices of *Brassica napus* Plants During Vernalization, *Plant Sci.* 2012, *187*, 105–112.

[221] Moravcova, E., Vaclavik, L., Lacina, O., Hrbek,V., Riddellov, K., Hajslov, J. Novel Approaches to Analysis of 3-chloropropane-1,2-diol Esters in Vegetable Oils, *Anal. Bioanal. Chem.* 2012, *402*, 2871–2883.

[222] Mimmo T., Hann S., Jaitz L., Cesco S., Gessa C., Puschenreiter M., Time and Substrate Dependent Exudation of Carboxylates by *Lupinus albus L.* and *Brassica napus L.*, *Plant Physiol. Biochem.* 2011, *49*, 1272–1278.

[223] Jaitz, L., Mueller, B., Koellensperger, G., Huber, D., Oburger, E., Puschenreiter, M., Hann, S. LC–MS Analysis of Low Molecular Weight Organic Acids Derived from Root Exudation, *Anal. Bioanal. Chem.* 2011, *400*, 2587–2596.

[224] Bouali, I., Trabelsi, H., Herchi, W., Martine, L., Albouchi, A., Bouzaien, G., Sifi, S., Boukhchina, S., Berdeaux, O. Analysis of Pecan Nut (*Carya illinoinensis*) Unsaponifiable Fraction. Effect of Ripening Stage on Phytosterols and Phytostanols Composition, *Food Chem.* 2014, *164*, 309–316.

[225] Jeong, T., Itoh, T., Tamura, T., Matsumoto, M. Analysis of Sterol Fractions from Twenty Vegetable Oils, *Lipids* 1974, *9*, 921–927.

[226] Zhang, L., Liu, C., Meng, Y. Isolation and Determination of Sterols in Rapeseed Oil by HPLC-MS, *Anal. Letts*. 2009, *42*, 1650–1661.

[227] Gwatidzo, L., Botha, P., McCrindle, R., Combrinck, S. Extraction and Identification of Phytosterols in Manketti (*Schinziophyton rautanenii*) Nut Oil, *J. Am. Oil Chem. Soc*. 2014, *91*, 783–794.

[228] Schurz, H., Hill, R., Philen, R., De la Paz, M.: Abaitua Borda, I.: Kilbourne, E., Bernert, T., Needham, L. Analytical Measurements of Products of Aniline and Triglycerides in Oil Samples Associated with the Toxic Oil Syndrome, *Arch. Toxic*. 1997, *19*, 53–64.

[229] Crow, F., Cragun, J., Johnson, K., Ruiz, M., De La Paz, M., Naylor, S. On-line HPLC-UV-mass spectrometry and Tandem Mass Spectrometry for the Rapid Delineation and Characterization of Differences in Complex Mixtures: A Case Study Using Toxic Oil Variants, *Biomed. Chromatogr*. 2002, *16*, 311–318.

[230] Kammuller, M., Verhaar, H., Versluis, C., Terlouw, J., Brandsma, L., Penniks, A., Seinen, W. 1-Pphenyl-5-vinyl-2-imidazolidinethione, a proposed causative agent of spanish toxic oil syndrome: synthesis and identification in one of a group of case-associated oil sample, *Fd. Chem. Toxic.* 1988, *26*, 119–127.

[231] Bernert, J., Jr., Pendergrast, A., Ashley, D., Patterson, D., Jr., Kilbourne, E., Alexander, L. Synthesis of N-(5-vinyl-1,3-thiazoliudin-2-ylidene) phenylamine and analysis of oils implicated in the spanish toxic syndrome for its presence, *Fd. Chem. Toxic*. 1989, *27*, 159–164.

[232] Yoshida, S., Bruenner, B., German, J., Gershwin, M. Decrease of 12-Hydroxyeicosatetraenoic Acid Production in Mouse Lungs Following Dietary Oleic Anilide Consumption: Implications for the Toxic Oil Syndrome, *Arch. Environ. Contain. Toxicol.* 1995, *28*, 524–528.

[233] Hill, R., Jr., Schurz, H., De la Paz, M., Borda, I., Philen, R., Kilbourne, E., Head, S., Bailey, S., Driskell, W., Barr, J., Needham, L. Possible Etiologic Agents for Toxic Oil Syndrome: Fatty Acid Esters of 3-(N-Phenylamino)-l,2-Propanediol, *Arch. Environ. Contam. Toxicol.* 1995, *28*, 259–264.

[234] Guitart, R., Abian, J., Arboix, M., Gelp, E. Detection and Isolation of Pentachlorophenol in Oil Samples Associated with the Spanish Toxic Oil Syndrome, *Bull. Environ. Contam. Toxicol.* 1990, *45*, 181–188.

[235] Quero, C., Colome, N., Rodriguez, C., Eichhorn, P., De la Paz, P., Gelpi, E., Abian, J. Proteomics of Toxic Oil Syndrome in Humans: Phenotype

Distribution in a Population of Patients, *Chem.-Biol. Inter.* 2011, *192*, 129–135.
[236] Morato, A., Escabros, J., Manich, A., Reig, N., Castano, Y., Abian, J., Messeguer, A. On the Generation and Outcome of 3-(N-Phenylamino)propane-1,2-diol Derivatives in Deodorized Model Oils Related to Toxic Oil Syndrome, *Chem. Res. Toxicol.* 2005, *18*, 665–674.
[237] Jiang, Y., Li, Y., Jiang, Y., Li, J., Pan, C. Determination of Multiresidues in Rapeseed, Rapeseed Oil, and Rapeseed Meal by Acetonitrile Extraction, Low-Temperature Cleanup, and Detection by Liquid Chromatography with Tandem Mass Spectrometry, *J. Agric. Food Chem. 2012*, 60, 5089–5098.
[238] Walorczyk, S., Gas chromatographic–tandem Mass Spectrometric Analysis of Pesticides Residues in Produce Using Concurrent Solvent Recondensation-large Volume Injection, *J. Chromatogr. A* 2012, *1222*, 98–108.
[239] Ioannidou, O., Zabaniotou, A., Stavropoulos, G., Islam, M., Albanis, T. Preparation of Activated Carbons From Agricultural Residues for Pesticide Sdsorption, *Chemosphere* 2010, *80*, 1328–1336.
[240] Guan, W., Li, Z., Zhang, H., Hong, H., Rebeyev, N., Ye, Y., Ma, Y. Amine Modified Graphene as Reversed-dispersive Solid Phase Extraction Materials Combined with Liquid Chromatography-tandem Mass Spectrometry for Pesticide Multi-residue Analysis in Oil Crops, *J. Chromatogr. A* 2013, *1286*, 1–8.
[241] You X., Liang L., Liu F., Dissipation and Residues of Clethodim and its Oxidation Metabolitesin a Rape-field Ecosystem Using QuEChERS and Liquid Chromatography/tandem Mass Spectrometry, *Food Chem.* 2014, *143*, 170–174.
[242] Ivanova, B., Spiteller, M. Simultaneous Quantitation of Naturally Occurring Insecticides, Acaricides, and Piscicides in Rapeseed Oil by UV-MALDI Mass Spectrometry, *J. Food Measur. Character.* 2014, *8*, 15–28.
[243] Regulation (EC) No 396/2005, European Commission, http://ec.europa.eu/sanco_pesticides/public/(May 2011).
[244] Directive and further amendments. Off. J. Eur. Union L350/14.12.1990 and 1990/642/EC.
[245] Verordnung (EG) Nr. 1272/2008 Eur. Parlaments und des Rates (16.12.2008).
[246] Verordnung (EU) Nr. 618/2012 der Kommission (10.07.2012).

[247] Kara, D., Fisher, A., Hill, S. Extraction of Trace Elements by Ultrasound-assisted Emulsification from Edible Oils Producing Detergentless Microemulsions, *Food Chem.* 2015, *188*, 143–148.

[248] Eggert, K., Von Wiren, N. Dynamics and Partitioning of the Ionome in Seeds and Germinating Seedlings of Winter Oilseed Rape, *Metallom.* 2013, *5*, 1316–1325.

[249] Ouerdane, L., Aureli, F., Flis, P., Bierla, K., Preud'homme, H., Cubadda, F., Szpunar, J. Comprehensive speciation of low-molecular weight-selenium metabolites in mustard seeds using HPLC – electrospray linear trap/orbitrap tandemmass spectrometry, *Metallom.* 2013, *5*, 1294–1304.

[250] Aubert, T., Burel, A., Esnault, M., Cordier, S., Grasset, F., Cabello-Hurtado, F. Root Uptake and Phytotoxicity of Nanosized Molybdenum Octahedral Clusters, *J. Hazardous Mater.* 2012, *219–220*, 111–118.

[251] Pavlickova, J., Zbiral, J., Smatanova, M., Houserova, P., Cizimarova, E., Houserova, S., Kuban, V. Uptake of Thallium from Artificially and Naturally Contaminated Soils into Rape (*Brassica napus L.*), *J. Agric. Food Chem.* 2005, *53*, 2867–2871.

[252] Xu, B., Zhang, L., Wang, H., Luoi, D., Li, P. Characterization and Authentication of Four Important Edible Oils Using Free Phytosterol Profiles Established by GC–GC–TOF/MS, *Anal. Met.* 2014, *6,* 6860–6870.

[253] Hrbek, V., Vaclavik, L., Elich, O., Hajslova, J. Authentication of Milk and Milk-based Foods by Direct Analysis in Real Time Ionizationehigh Resolution Mass Spectrometry (DART–HRMS) Technique: A Critical Assessment, *Food Control* 2014, *36*, 138–145.

[254] Catharino, R., Haddad, R., Cabrini, L., Cunha, I., Sawaya, A., Eberlin, M. Characterization of Vegetable Oils by Electrospray Ionization Mass Spectrometry Fingerprinting: Classification, Quality, Adulteration, and Aging, *Anal. Chem.* 2005, *77*, 7429–7433.

[255] Sghaier L., Cordella C., Rutledge D., Watiez M., Breton S., Kopczuk A., Sassiat P., Thiebaut D., Vial J., Comprehensive Two–dimensional Gas Chromatographyfor Analysis of the Volatile Compounds and Fishy Odor Off–flavors from Heated Rapeseed Oil, *Chromatographia* DOI 10.1007/s10337-015-2897-8, 2015.

[256] Greer, M., Zhou, T., Weselake, R. A Novel Assay of DGAT Activity Based on High TemperatureGC/MS of Triacylglycerol, *Lipids* 2014, *49*, 831–838.

[257] Granvog, M. Development of Three Stable Isotope Dilution Assays for theQuantitation of (E)-2-Butenal (Crotonaldehyde) in Heat-

ProcessedEdible Fats and Oils as Well as in Food, J. Agric. Food Chem. 2014, *62*, 1272–1282.

[258] Ewert, A., Granvog, M., Schieberle, P. Development of Two Stable Isotope Dilution Assays for the Quantitation of Acrolein in Heat-Processed Fats, *J. Agric. Food Chem.* 2011, *59*, 3582–3589.

[259] Tarvainen, M., Suomela, J., Kuksis, A., Kallio, H. Liquid Chromatography–Light Scattering Detector–Mass Spectrometric Analysis of Digested Oxidized Rapeseed Oil, *Lipids* 2010, *45*, 1061–1079.

[260] Jelen, H., Mildner-Szkudlarz, S., Jasinska, I., Wasowicz, E. A Headspace–SPME–MS Method for Monitoring Rapeseed Oil Autoxidation, *J. Am. Oil Chem. Soc.* 2007, *84*, 509–517.

[261] Hong, E., Son, H., Choi, J.: Noh, B. Authentication of Rapeseed Oil Using an Electronic Nose Based on Mass Spectrometry, *Kor. J. Food Sci. Technol.* 2011, *43*, 105–109.

[262] Hu, W., Zhang, L., Li, P., Wang, X., Zhang, Q., Xu, B., Sun, X., Ma, F., Ding, X. Characterization of Volatile Components in Four Vegetable Oils by Headspace Two-Dimensional Comprehensive Chromatography Time-of-flight Mass Spectrometry, *Talanta* 2014, *129*, 629–635.

[263] Chardin, H., Mayer, C., Senechal, H., Poncet, P., Clement, G., Wal, J., Desvaux, F., Peltre, G. Polygalacturonase (*pectinase*), a New Oilseed Rape Allergen, *Allergy: Eur. J. Allergy Clin. Immunol.* 2003, *58*, 407–411.

[264] Yuan, L., Guo, W., Wang, Z. The extraction, Separation and Structural Ascertainment of 4-pentenyl Glucosinolates in Rapeseed, *Fenxi Huaxue* 2004, *32*, 1513–1516.

[265] Zhou, J., Hu, J., Qiu, A. Separation and Purification of the Main Glucosinolate from Rapeseeds, *Ch. J. Chromatogr.* 2005, *23*, 411–414.

[266] Pustjens, A., Schols, H., Kabel, M., Gruppen, H. Characterisation of Cell Wall Polysaccharides from Rapeseed (*Brassica napus*) meal, *Carbohydr. Polym.* 2013, *98*, 1650–1656.

In: Rapeseed
Editor: Monica White

ISBN: 978-1-63484-227-3

Chapter 5

ANTIOXIDANT ACTIVITY IN TRANSGENIC CANOLA (*BRASSICA NAPUS* L.) PLANTS GROWN *IN VITRO*

Liudmyla O. Sakhno*, Kateryna V. Lystvan, Valeriia B. Belokurova and Mykola V. Kuchuk
Institute of Cell Biology and Genetic Engineering NAS of Ukraine, Kyiv, Ukraine

ABSTRACT

The research was aimed at studying the possible unintended biochemical peculiarities of the recently obtained herbicide-resistant canola (*Brassica napus* L.) plants which express both the *BAR* gene conferring resistance to glufosinate herbicides and synthetic *EPSPS* gene responsible for resistance to glyphosates. Both total radical scavenging activity and superoxide dismutase activity as a part of plant antioxidant system were investigated in the leaf extracts using DPPH (2,2-diphenyl-1-picrylhydrazyl radical) and nitroblue tetrazolium assays, respectively. The total soluble protein content was also evaluated. Significant decrease in the total radical scavenging activity was detected for the primary transgenic plants possessing two different genes for herbicide resistance.

* Corresponding author: Liudmyla O. Sakhno, Department of Genetic Engineering, Institute of Cell Biology and Genetic Engineering of National Academy of Sciences of Ukraine, 148 Academika Zabolotnoho Str., 03143, Kyiv, Ukraine, e-mail: sakhno@icbge.org.ua, sakhno2007@ukr.net.

The total soluble protein content and superoxide dismutase activity were unaffected in these plants. The analyzed biochemical parameters in the leaf extracts from untransformed plants were similar to those of the herbicide-resistant plants of the first and the second generations grown *in vitro* on the media without herbicide supplementation. Addition of Basta (phosphinothricin) or Hurricane Forte (glyphosate) herbicide solutions to the culture media did not affect antioxidant activity of transgenic canola plants. Thus, T_2 generation homozygous canola plants resistant to both glufosinate and glyphosate herbicides had the total soluble protein content, total radical scavenging activity, and superoxide dismutase activity in their leaves similar to that of the untransformed control plants. Earlier we have produced several groups of transgenic canola lines bearing target genes of the different origins. There were the plants expressing either 1) both DesC desaturase (*Des*C) gene from cyanobacterium *Synechococcus vulcanus* and *EPSPS* gene or 2) human interferon α2b (*HuIFN*-α2b) gene or 3) *CYP11A1* gene for cytochrome $P450_{SCC}$ from bovine adrenal cortex mitochondria. Antioxidant activities in these plants have been compared to that of the newly obtained herbicide-resistant plants. The influence of different target genes on plant antioxidant activity and the importance of this parameter for both plants and human health are discussed.

Keywords: *Brassica napus,* antioxidant activity, total radical scavenging activity, superoxide dismutase

INTRODUCTION

Herbicide-resistant varieties have consistently occupied the largest planting area among biotech crops since their advent in 1996. In 2014 both glufosinate- and glyphosate-resistant rapeseed cultivars were grown on 9 million hectars, what was 25% of *the* total planting area *for this crop worldwide* (James, 2014).

Recently, a new tendency in the design of herbicide-resistant varieties emerged. It is the combination of resistance to several herbicides simultaneously (Vencil et al., 2012). Thus, Triazine resistant Roundup Ready canola plants were produced through a hybrid breeding techniques by Monsanto Company and Pacific Seeds worked in conjunction. They possess resistance to glyphosate and triazins (http://roundupreadycanola.com.au/wp-content/uploads/2014/03/What-is-Triazine-Tolerant-Roundup-Ready-canola1.pdf).

Although there is a number of transgenic plant peculiarity investigations (Reguera et al., 2012; Bhargava and Sawant, 2013), the discussions about their potential environmental risks and danger are continued. Changes both in total radical scavenging and superoxide dismutase (SOD) activities as the measures of antioxidant activity in transgenic plants may cause great influence on plant development. Antioxidant activity of plant tissues characterizes an ability to maintain the balance of reactive oxygen species (ROS) needed for plant metabolism (Karuppanapandian et al., 2011, Mittlera and Blumwald, 2015). ROS includes both the free radicals such as superoxide, peroxyl and hydroxyl radicals and the molecules such as singlet oxygen and hydrogen peroxide. ROS are commonly produced during photosynthesis, respiration, photorespiration, and primary as a result of a membrane-bound-NADPH activity (De Gara et al., 2003; Pilon et al., 2011; Marino et al., 2012). High light, extreme temperatures, drought, flooding, salt stress, pathogen attack, and especially the combinations of various stresses dramatically increase a quantity of these substances and may lead to oxidative damage (Susuki et al., 2011; Pandey et al., 2014). Plant antioxidant system is responsible for overcoming the excessive amounts of ROS. It provides the tightly coordinated activity of enzymes (SOD, peroxidase, catalase) and non-enzyme antioxidants (proline, glutathione, ascorbic acid, fatty acids, polyphenols (tocopherols, anthocyanins, flavonoids, aromatic oxyacids)) (Hasanuzzaman et al., 2012, Wojtera-Kwiczor et al., 2014). The major nectar protein from ornamental tobacco is superoxide dismutase that functions to generate high levels of hydrogen peroxide against pathogenic fungi and bacteria (Thornburg et al., 2003). The positive correlation of treatments using the high functional substances and stress tolerance improvement due to increase in antioxidant activity was also proved (Liu et al., 2005; Chung et al., 2006; Janeczko 2012). Changes in antioxidant activity in plants of the same species grown in different environmental conditions were noticed (Abbasi et al., 2015). SOD activity in the alpine plants was higher than in plants grown in the steppe (Öncel et al., 2004). Positive correlation between activities of antioxidant enzymes (SOD, catalase and glutathione peroxidise) and a radical scavenging activity detected using DPPH assay was found in Chinese medicine plants (Lee et al., 2003). Some compounds, for example, coronatine which is a chlorosis-inducing phytotoxin that mimics some biological activities of methyl jasmonate may alleviate plant stress influence by increase in both SOD and radical scavenging activities (Xie et al., 2008). Increase in antioxidant activity due to transgenesis positively affects the plant abilities to grow faster and to resist the stresses of different origins (Gusta et al., 2009; Luo et al., 2013; Confalonieri et al.,

2014). Values of SOD activity were used for *in vitro* salt tolerance evaluation of sweet potato (*Ipomoea batatas* (L.) Lam.) mutants (He et al., 2009). This approach facilitated the following selection of salt tolerant plants under greenhouse conditions. Tissues with lower amount of antioxidants including SOD had decreased responsiveness to *in vitro* manipulations and genetic transformation (Dan, 2008). SOD activity was higher in *in vitro* regenerated and propagated *Leptadenia reticulate* plants as compared to field grown ones (Sudipta et al., 2014).

SOD activity was studied in seeds of herbicide-resistant canola and maize (*Zea mays* L.) (Xu et al., 2011). Earlier we have produced transgenic canola plants which express both *BAR* and synthetic *EPSPS* gene and determined SOD activity in their leaf tissue in non-stress conditions (Sakhno and Slyvets 2014). However, this parameter has not yet been explored under stress (herbicide application). In addition, any data about the total radical scavenging activity in herbicide-resistant plants are absent. Study of these characteristics will allow determining the possible unintended biochemical changes in transgenic canola in comparison with non-transgenic plants.

METHODS

Plant material and growth conditions. Biotech canola plants expressing both *BAR* and *EPSPS* transgenes have been produced as a result of *Agrobacterium tumefaciens*-mediated transformation on the cv Exgold background by using pCB133 vector (Figure 1) according to our own protocol (Sakhno et al., 2008). Primary transformants were acclimatized to greenhouse conditions and self-pollinated. The produced seeds (T_1 generation) were sterilized and tested for phosphinothricin and glyphosate resistance *in vitro*. The same procedure was held for T_2 seed production. The best lines according to glyphosate resistance under greenhouse conditions, namely $T_0$15/133/2 (primary transformant), $T_1$15/133/2/9 (T_1 generation), and $T_2$15/133/2/9 (T_2 generation), were chosen for the current analyses. The second generation line was characterized to be homozygous (Sakhno et al., 2015). Untransformed spring canola plants cv Exgold were used as a control. Both untransformed and biotech plants were micropropagated and grown *in vitro* (+23°C, 4000–5000 lx, 14-h photoperiod) on agar-solidified hormone-free MS medium (Murashige and Skoog 1962) in Magenta^{TM} Boxes for four weeks. The upper and adventitious buds of both $T_2$15/133/2/9 and untransformed lines were also cultured together on the same media supplemented with either phosphinothricin (5 mg/L) or glyphosate (2 mg/L). The sterile herbicide solutions were added to the media after autoclaving.

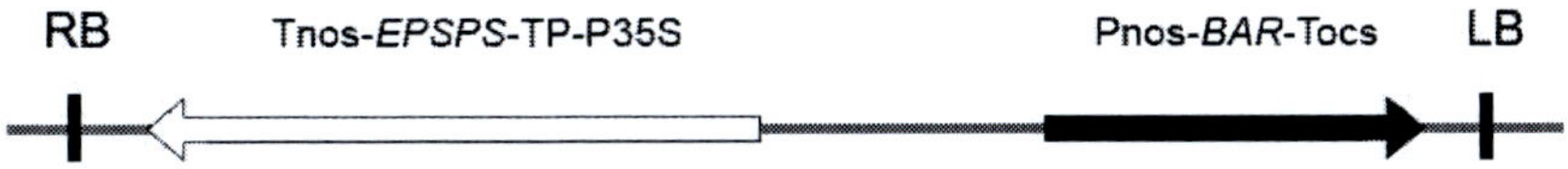

Figure 1. Scheme of the T-DNA region of plasmid construct used in the experiments: LB – T-DNA left border; RB – T-DNA right border; Pnos-*BAR*-Tosc – phosphinothricin acetyltransferase (*BAR*) gene expression cassette (coding sequence for BAR with nopalin synthase gene promoter (Pnos) and octopine synthase gene terminator (Tocs); P35S-TP-*EPSPS*-Tnos – 5-enolpyruvylshikimate-3-phosphate synthase gene (*EPSPS*) expression cassette (fusion of N-terminal chloroplast transit peptide (TP) with synthetic ESPSP coding sequence downstream of cauliflower mosaic virus 35S promoter (P35S) and followed by nopalin synthase gene terminator (Tnos).

Determination of Total Soluble Protein (TSP) Content

TSP was determined according to Bradford (Bradford 1976). The fresh plant material (100 mg) was ground with 1 mL of 50 mM Tris-HCl buffer (pH 8.0) in a mixer mill Retsch MM 400 (Germany) with vibration frequency of 25Hz for 3 min and then centrifuged at 13000 g (4°C) for 15 min. The supernatant was used for the analyses. The optical density of the reaction mixture was detected at 595 nm by BioPhotomether Eppendorf, v.1.35 (Germany) with bovine serum albumin (Fermentas, Lithuania) as a standard.

SOD Activity Assay

SOD activity was measured using nitroblue tetrazolium photochemical oxidation method (Beyer and Fridovich, 1987). Formazan formation was held in 1.5-mL Eppendorf tubes. Plant extracts could inhibit this reaction due to the SOD activity. One tube for each sample was retained in the dark. Three others were illuminated with white light (fluorescent lamp T5/G5, model ELI-230A-T5-8W) for 5 min in the thermostat at 24°C. The color of the reaction mixture changed from yellow to violet as the result of oxidation. In the null sample containing no leaf extract the oxidation was complete. The reaction mixture absorbance in the illuminated samples was determined at 550 nm by BioPhotomether Eppendorf (Germany) and compared with corresponding values of dark samples.

Total Radical Scavenging Activity Assay

DPPH (2,2-diphenyl-1-picrylhydrazyl) assay was used for the total radical scavenging activity evaluation (Blois 1958). Reactions were conducted in 96-well microplates. Each well contained 100 µL of the reaction mixture consisting of DPPH solution in 96% ethanol (10^{-4} M) and 20; 10; 5; 2.5; 1.25 or 0.625 µL of the tested extract. The reaction was carried out for 30 min in the dark at room temperature. Then, the optical densities of the final solutions were measured at 550 nm. The optical density of plant extract solutions in ethanol without DPPH addition was also determined. Ascorbic acid solution in water (10 mg/mL) was used as the positive control. DPPH reduction was expressed as a percent of inhibition and calculated according to the following formula:

$$A(c) - A(e) / A(c) * 100,$$

where A(c) was the optical density of the control solution, and

A(e) was the optical density of solution of the tested extract.

The correlation graph was calculated on the basis of values of DPPH color inhibition and the extract volume needed for it. The volume of extract that caused 50% inhibition was determined as EC_{50} value.

Statistical Analysis

The experiments were repeated three times independently, and each data point was the mean of three replicates. Average mean, standard deviation, confidence interval and Student's t-test for statistical significance were calculated using Microsoft ® Office Excel 2003 (Microsoft Corporation) standard functions.

RESULTS AND DISCUSSION

The initial untransformed and biotechnological canola lines bearing both *EPSPS* and *BAR* transgenes were cultured *in vitro*. The upper and adventitious buds were grown on agar-solidified hormone-free MS medium either with herbicide addition or without it (Figure 2).

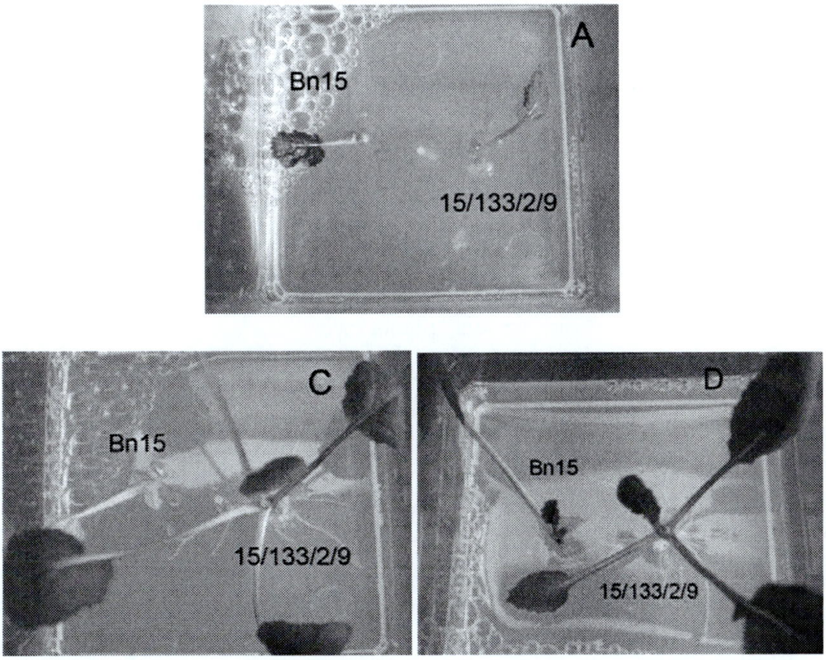

Figure 2. The adventitious canola buds growth on the media supplemented with phosphinothricin (A, C) and glyphosate (B, D). A, B – the start of the experiment, B, D – 28 day of the experiment, Bn15 – control, untransformed plants cv Exgold, 15/133/2/9 – T_2 generation transgenic line.

When the control buds were cultured on the medium supplemented with phosphonomethylglycine (glyphosate) for four weeks (2 mg/L, the Hurricane Forte causative agent), they had no ability to form the roots and produced the only small leaf approximately 2 mm in diameter (Figure 2, D). Under the same conditions the adventitious transgenic buds of $T_2$15/133/2/9 line have developed into plants with the powerful root system and leaves (Figure 2, D). The effect of phosphinothricin (5 mg/L) on the control plant growth was stronger in comparison to that of glyphosate (Figure 2). Untransformed adventitious buds bleached and died, nevertheless the transgenic ones produced normal shoots with roots and leaves under phosphinothricin pressure (Figure 2, C).

TSP content is one of the parameters that are used for evaluation of plant physiological state both in normal and stress conditions. Thus, water deficit caused the increase in TSP content in tomato (*Lycopersicon esculentum*) (Mäkelä et al., 2000) and sunflower (*Helianthus annuus* L.) (Iqbal et al., 2011) leaves. In contrast, the same conditions led to decrease in the TSP content in sesame (*Sesamum indicum* L.) cultivars (Fazelli et al., 2007). It was noticed that plants of two sesame varieties which were different in TSP under physiological conditions demonstrated different response to drought. Decrease in TSP was lower under water deficit in the cultivar that was characterized by higher drought tolerance and higher TSP content in normal conditions (Fazelli et al., 2007). Protein concentration may be influenced by some treatments, for example it increased in 28-homobrassinolide-treated maize plants (Arora et al., 2008).

Leaves of 28-day-old canola plants were used for the biochemical analyses in our experiments. TSP determination showed no significant differences between the untransformed (24.15 ± 1.3 mg/g fresh weight (FW)) and biotechnological plants of different generations grown in physiological conditions (23.85 ± 1.1; 24.2 ± 0.9 and 23.86 ± 1.5 mg/g FW in $T_0$15/133/2; $T_1$15/133/2/9 and $T_2$15/133/2/9 respectively) (Figure 3). Both glyphosate and phosphinothricin addition to culture media did not influence TSP content in $T_2$15/133/2/9 plants (Figure 3). Although in non-stress conditions the similar results were demonstrated for transgenic canola plants expressing either DesC desaturase (*Des*C) gene from cyanobacterium *Synechococcus vulcanus* simultaneously with *EPSPS* gene or human interferon α2b (*HuIFN*-α2b) gene (Slyvets et al., 2013; Sakhno and Slyvets, 2014, Sakhno, 2015), in the latter plant group the TSP content was higher as compared to the controls under osmotic stress induced by mannitol (Slyvets and Sakhno, 2014). In addition, drought tolerance improvement in *HuIFN*-α2b canola plants was shown in terms of fresh weight, membrane permeability, photosynthetic pigment content, and SOD activity. In contrast, no alterations were determined in the TSP content for *Des*C/*EPSPS* canola lines in comparison with wild-type plants both in cold and osmotic stresses (Slyvets et al., 2013; Sakhno 2015) as well as no increase in tolerance to these stresses was confirmed. Earlier we have also produced canola plants bearing *CYP11A1* gene for cytochrome $P450_{SCC}$ from bovine adrenal cortex mitochondria (Sakhno et al., 2010). TSP content of these lines was significantly higher (up to 1.36-fold) than that of the initial plants under non-stress conditions, but it decreased to the level of the wild-type plants under osmotic stress (Sakhno 2015). All transgenic canola plants mentioned above were resistant to glufosinate herbicide due to the expression of *BAR* gene which was used in transformation vectors as a selective one. It allows concluding that *BAR* gene expression did not affect TSP content in transgenic canola plants. Moreover, biotechnological canola plants expressing either *BAR*/*EPSPS* or *BAR*/*EPSPS*/*Des*C genes had TSP content similar to untransformed controls. Thus, *EPSPS* gene expression did not lead to change in leaf TSP content.

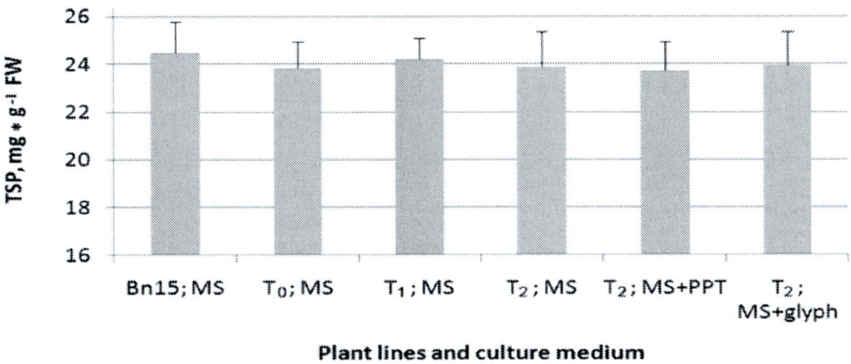

Figure 3. Leaf TSP content in canola plants grown *in vitro*: Bn15 – control untransformed plants cv Exgold; T_0 – $T_0$15/133/2 (the primary transformant); T_1 – $T_1$15/133/2/9 (T_1 generation transgenic line), and T_2 – $T_2$15/133/2/9 (T_2 generation transgenic line), PPT – phosphinothricin; glyph – glyphosate. Error bar represents mean ± one standard deviation from three replications (n = 3).

The total radical scavenging activity evaluation in the leaf extracts of the analyzed canola plants demonstrated no significant differences between untransformed and biotechnological ones of different generations except the primary transgenic $T_0$15/133/2 line (Figure 4). The scavenging activity against DPPH radical in these plants was lower by 44.5% as compared to wild-type plants. One of the possible explanations may be a change in redox homeostasis because of hemizygous status of the target genes for herbicide resistance. In the next generations it might be levelled due to transition of these genes to homozygous state. Moreover, the total radical scavenging activity in the leaves of transgenic $T_2$15/133/2/9 line (the second generation) cultured on neither the media with phosphinothricin or glyphosate did not differ significantly from values for the same media without herbicides (Figure 4). It evidenced no stress response caused by corresponding heterologous gene expression in the analyzed plants. In contrast, increase in antioxidant activity in the leaf extracts of *CYP11A1* canola by 38–130% was testified (Sakhno et al., 2010) using 2,6-Dichloroindophenol, Sodium salt oxidation method (Semenov and Yarosh, 1985). In addition, these canola as well as tobacco plants possessing the same transgene were characterized by increase in biomass production, TSP content, and growth rate (Spivak et al., 2009; Sakhno et al., 2010). *CYP11A1* canola plants also demonstrated higher adaptive plasticity in short time heat (Sakhno et al., 2014) and osmotic stresses (Sakhno 2015) in comparison with untransformed ones. Increment in antioxidant activity up to 4-fold was also detected in leaves of *PAP1* (Production of Anthocyanin Pigment 1) canola (Li et al., 2010). Both tobacco and canola plants bearing *CYP11A1* gene as well as *PAP1* canola had *BAR* gene in their nuclear genome. Comparison of changes in antioxidant activity of these plants and the analyzed in the present work resulted to the conclusion that *CYP11A1* and *PAP1* gene expression influenced the total radical scavenging activity while *BAR* gene expression did not.

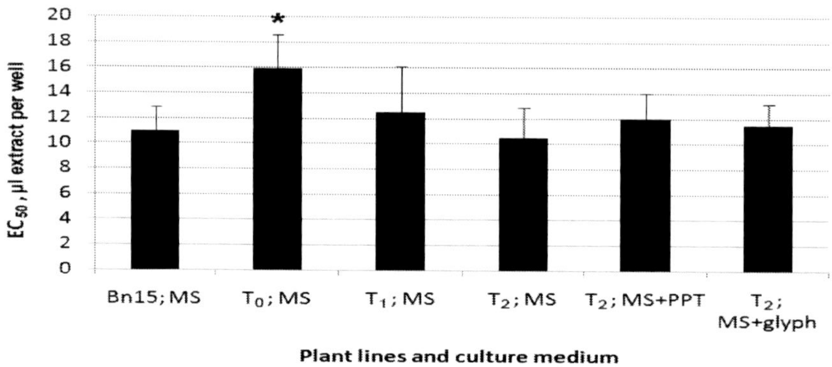

Figure 4. DPPH in the leaf extracts of canola plants grown *in vitro*: Bn15 – control untransformed plants cv Exgold; T_0 – $T_0$15/133/2 (the primary transformant); T_1 – $T_1$15/133/2/9 (T_1 generation transgenic line), and T_2 – $T_2$15/133/2/9 (T_2 generation transgenic line), PPT – phosphinothricin; glyph – glyphosate. Error bar represents mean ± one standard deviation from three replications (n = 3). *Asterisk* * indicates significant difference of the experimental value as compared to that of the control one ($p \leq 0.05$).

SOD activity may be used as an indicator of the redox homeostasis in plants and animals. This parameter changes at the different plant development stages (Gupta et al., 1993; Diaz-Vivancos et al., 2010; Matamoros et al., 2010). SOD activity differences were detected in different organs of the same plant (Martins et al., 2011). The stress stimuli of both biotic and abiotic origins cause the strong influences on plant SOD activity (Molina-Rueda et al., 2013).

SOD activity measurements in leaf extracts of both untransformed and *BAR/EPSPS* canola revealed no significant differences between the wild-type and transgenic plants (Figure 5) grown on MS medium. Similarly, the means of SOD activity of biotechnological plants grown on the media with either phosphinothricin (5 mg/L; 22.4 ± 0.9 U mg^{-1} protein) or glyphosate (2 mg/L;

23.4 ± 2.0 U mg^{-1} protein) did not significantly differ as compared to that of the same plants cultured without herbicides (21.9 ± 1.6 U mg^{-1} protein).

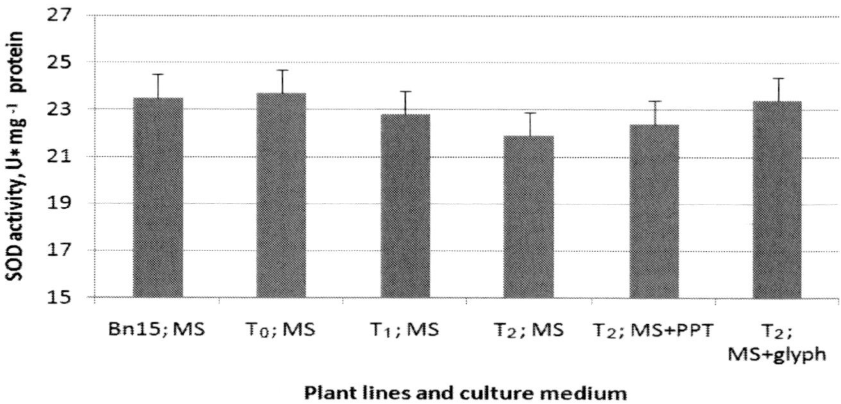

Figure 5. Leaf SOD activity in canola plants grown *in vitro*: Bn15 – control untransformed plants cv Exgold; T_0 – $T_0$15/133/2 (the primary transformant); T_1 – $T_1$15/133/2/9 (T_1 generation transgenic line), and T_2 – $T_2$15/133/2/9 (T_2 generation transgenic line), PPT – phosphinothricin; glyph – glyphosate. Error bar represents mean ± one standard deviation from three replications (n = 3).

The SOD activity evaluation in seeds of the commercial transgenic canola events, namely GT73, MS1/RF1 and MS8/RF3 testified decrease in this parameter by 22% and 25% for the first and second plant group mentioned, respectively, and no change for the latter one in comparison with nontransgenic controls (Xu et al., 2011). The plants of GT73 event is characterized by glyphosate resistance due to *EPSPS* gene expression. Both MS1/RF1 and MS8/RF3 canola plants possess *BAR* gene in their nuclear genome and are resistant to glufosinate treatment. Thus, either the *BAR* or *EPSPS* gene expression may influence SOD activity in canola seeds. However, no alterations in leaf canola SOD activity were found in our experiments with *BAR*/*EPSPS* plants (Figure 5). It was similar to the results obtained for MS8/RF3 canola seeds (Xu et al., 2011). In contrast, SOD activity rise was determined in both *HuIFN*-α2b and *CYP11A1* canola leaves under the normal conditions (Sakhno and Slyvets, 2014) as well as under water deficit (Slyvets and Sakhno, 2014, Sakhno, 2015) and heat (Sakhno et al., 2014). It was accompanied by higher biomass and TSP production as well as photosynthetic pigment content and lower membrane permeability in transgenic plants under stresses. A lot of the other transgenic plants characterized by the SOD activity increase in normal conditions due to heterologous gene expression have been produced by the top world laboratories (Gusta et al., 2009; Confalonieri et al., 2014; Su et al., 2015) in an effort to elevate some crop biomass and yield under the climate changes. However, majority of transgenic plants are similar in term of SOD activity as compared to untransformed initial plants under physiological conditions. Oftentimes the ability to higher SOD activity increase in comparison to nontransgenic controls was documented for them under stresses (Qiao et al., 2010; Lee et al., 2013; Bai et al., 2013). So, the innate elevated SOD activity in some wild plants and crop varieties as well as the acquired one due to transgenesis gives the advantages for plant growth and development especially in nonphysiological conditions.

The correlation between the radical scavenging and SOD activities in plants was noticed. Thus, the elevated SOD activity was proportional to the stimulation of antioxidant activity in fava bean (*Vicia faba*) sprouts after a seed microwave treatment (Randhir and Shetty, 2004).

ROS are commonly produced during normal plant, animal and human metabolism. The excessive quantity of these compounds leads to oxidative damage. In humans, ROS are involved in pathogenesis of cancer, cardiovascular diseases, atherosclerosis, hypertension, ischemia/reperfusion injury, diabetes mellitus, neurodegenerative diseases (Alzheimer's disease and Parkinson's disease), rheumatoid arthritis, and ageing (Valko et al., 2007). In some cases, for example in hypertension and diabetes mellitus, a potential usefulness of plant antioxidants was confirmed (Galvez Ranilla et al., 2010; Akolade et al., 2014). There are the highest quantity of the antioxidants in fruits as compared to such groups of dietary products as berries; nuts, seeds and dry fruits; vegetables; roots and tubers; beans; cereals (Halvorsen et al., 2002). Plants used in the folk medicine of different world regions are characterized by high antioxidant activity. Thus, methanol extracts of some plants traditionally used in Chinese medicine demonstrated high level DPPH scavenging activity and could inhibit H_2O_2-induced apoptosis of Chinese hamster lung fibroblast (V79-4) cells (Lee et al., 2003). The extract of *Areca catechu* var. *dulcissima* possessing the highest DPPH value provided most significant dose-dependent increase of SOD, catalase and glutathione peroxidise activities in V79-4 cells and thus protected them against oxidative damage. Supplementation of the maize-soybean meal based diets with rosemary (*Rosmarinus officinalis*) and its volatile oil improved broiler meat quality. Moreover, growth performance was positively affected by the rosemary volatile oil supplementations (Yesilbag et al., 2011). At the same time the research efforts aiming to elucidate whether excessive formation of free radicals is a primary cause or a downstream consequence of tissue injury in plants and animals as well as to investigate the usefulness of plants as a source of antioxidants for human health go on (Valko et al., 2007, Dudonné et al., 2009; Stevenson and Lowe, 2009; Karuppanapandian et al., 2011).

Conclusion

Thus, TSP content, DPPH and SOD activities in the leaves of T_2 generation homozygous canola plants grown *in vitro* and resistant to both glufosinate and glyphosate detected no significant differences as compared to that of the untransformed plants. Similarly, antioxidant activity was the same in untransformed controls and in transgenic canola plants expressing both DesC desaturase (*Des*C) gene from cyanobacterium *Synechococcus vulcanus* and *EPSPS* gene. In contrast, both SOD and radical scavenging activities were increased in canola plants bearing either human interferon α2b gene or *CYP11A1* gene for cytochrome $P450_{SCC}$ from bovine adrenal cortex mitochondria under non-stress conditions. In addition, osmotic stress tolerance improvement was demonstrated for both *CYP11A1* and *HuIFN*-α2b canola plants. Although the introduction of *EPSPS, BAR* and *Des*C genes did not affect the antioxidant status of canola plants, expression of some target genes (*CYP11A1* and *HuIFN*-α2b) enhanced their antioxidant activity. The mechanisms of these alterations require further investigation.

References

Abbasi, A.M., Guo, X., Nazir, A., (2015). Preliminary assessment of phytochemical contents and antioxidant properties of *Pistacia integerrima* fruit. *Pak. J. Pharm. Sci.* 28, 1187-1194.

Akolade, J.O., Usman, L.A., Okereke, O.E., Muhammad, N.O., (2014). Antidiabetic potentials of essential oil extracted from the leaves of *Hoslundia opposita* Vahl. *J Med Food.* 17, 1122–1128.

Arora, N., Bhardwaj, R., Sharma, P., Arora, H.K., (2008). 28-Homobrassinolide alleviates oxidative stress in salt-treated maize (Zea mays L.) plants. *Braz. J. Plant Physiol.* 20,153-157.

Bai, X., Liu, J., Tang, L., Cai, H., Chen, M., Ji, W., Liu, Y., Zhu, Y., (2013). Overexpression of *GsCBRLK* from *Glycine soja* enhances tolerance to salt stress in transgenic alfalfa (*Medicago sativa*). *Functional Plant Biol.* 40, 1048–1056.

Beyer, W.F., Fridovich, I., (1987). Assaying for superoxide dismutase activity some large consequences of minor changes in conditions. *Anal Biochem.* 161, 559–566.

Bhargava, S., Sawant, K., (2013). Drought stress adaptation: metabolic adjustment and regulation of gene expression. *Plant Breed.* 132, 21–32.

Blois, M.S., (1958). Antioxidant determinations by the use of a stable free radical. *Nature.* 181(4617), 1199–1200.

Bradford, M.M., (1976). A rapid and sensitive method for the quantitation of microgram quantities of protein utilizing the principle of protein-dye binding. *Anal Biochem.* 72, 248–254.

Chung, I.M., Kim, J.J., Lim, J.D., Yu, C.Y., Kim, S.H., Hahn, S.J., (2006). Comparison of resveratrol, SOD activity, phenolic compounds and free amino acids in *Rehmannia glutinosa* under temperature and water stress. *Environ. Exp. Bot.* 56, 44–53.

Confalonieri, M., Faè, M., Balestrazzi, A., Donà, M., Macovei, A., Valassi, A., Giraffa, G., Carbonera, D., (2014). Enhanced osmotic stress tolerance in *Medicago truncatula* plants overexpressing the DNA repair gene MtTdp2a (tyrosyl-DNA phosphodiesterase 2). *Plant Cell Tiss Organ Cult.* 116,187–203.

Dan, Y., (2008). Biological functions of antioxidants in plant transformation. *In Vitro Cell. Develop. Biol. - Plant*, 44, 149-161.

De Gara, L., de Pinto, M.C. Tommasi, F., (2003).The antioxidant systems vis-à-vis reactive oxygen species during plant–pathogen interaction. *Plant Physiol. Biochem.* 41, 863–870.

Diaz-Vivancos, P., Barba-Espin, G., Clemente-Moreno, M.J., Hernandez, J.A., (2010). Characterization of the antioxidant system during the vegetative development of pea plants. *Biol. plantarum.* 54, 76-82.

Dudonné, S., Vitrac, X., Coutière, P., Woillez, M., Mérillon, J.-M., (2009). Comparative study of antioxidant properties and total phenolic content of 30 plant extracts of industrial interest using DPPH, ABTS, FRAP, SOD, and ORAC assays. *J. Agric. Food Chem.* 57, 1768–1774.

Fazelli, F., Ghorbanli, M., Niknam, V., (2007). Effect of drought on biomass, protein content, lipid peroxidation and antioxidant enzymes in two sesame cultivars. *Biol. Plantarum.* 51, 98-103.

Galvez Ranilla, L., Kwon, Y.-In, Apostolidis, E., Shetty K., (2010). Phenolic compounds, antioxidant activity and *in vitro* inhibitory potential against key enzymes relevant for hyperglycemia and hypertension of commonly used medicinal plants, herbs and spices in Latin America. *Bioresource Technol.* 101, 4676–4689.

Gupta, S.A., Webb, R.P., Holaday, A.S., Allen, R.D., (1993). Overexpression of superoxide dismutase protects plants from oxidative stress (Induction of ascorbate peroxidase in superoxide dismutase-overexpressing plants). *Plant Physiol.* 103, 1067-1073.

Gusta, L.V., Benning, N.T., Wu, G., Luo, X., Liu, X., Gusta, M.L., McHughen, A., (2009). Superoxide dismutase: an all-purpose gene for agri-biotechnology. *Mol. Breed.* 24, 103–115.

Halvorsen, B.L., Holte, K., Myhrstad, M.C.W., Barikmo, I., Hvattum, E., Remberg, S.F., Wold, A.B., Haffner, K., Baugerod, H., Andersen, L.F., Moskaug, J., Jacobs, D.R., Blomhoff, Jr., Blomhoff, R., (2002). A systematic screening of total antioxidants in dietary plants. *J Nutr.* 132, 461–471.

Hasanuzzaman, M., Hossain, A.M., Teixeira da Silva, J.A., Fujita, M., (2012). Plant response and tolerance to abiotic oxidative stress: antioxidant defense is a key factor. In: B. Venkateswarlu et al., (eds.). Crop stress and its management: perspectives and strategies. Springer Netherlands; p.261-315.

He, S., Han, Y., Wang, Y., Zhai, H., Liu, Q., (2009). In vitro selection and identification of sweetpotato (*Ipomoea batatas* (L.) Lam.) plants tolerant to NaCl. *Plant Cell Tiss. Organ Cult.*, 96, 69-74.

Iqbal, N., Ashraf, Y., Ashraf, M., (2011). Modulation of endogenous levels of some key organic metabolites by exogenous application of glycine betaine in drought stressed plants of sunflower (*Helianthus annuus* L.). *Plant Growth Regul.* 63, 7–12.

James, C., (2014). Global status of commercialized biotech/GM crops: 2014. ISAAA Brief No. 49. ISAAA: Ithaca, NY. Available from http://www.isaaa.org/resources/publications/pocketk/16/.

Janeczko, A., (2012). The presence and activity of progesterone in the plant kingdom. *Steroids*.77, 169–73.

Karuppanapandian, T., Moon, J.-C., Kim, C., Manoharan, K., Kim, W., (2011). Reactive oxygen species in plants: their generation, signal transduction, and scavenging mechanisms. *Aust. J. Crop Sci.* 5, 709-725.

Lee, S.E., Hwang, H.J., Ha, J.S., Jeong, H.-S., Kim, J.H., (2003). Screening of medicinal plant extracts for antioxidant activity. *Life Sci.* 73, 167–179.

Lee, Y.P., Ahmad, R., Lee, H.S., Kwak, S.S., Kwon, S.Y., (2013). Improved tolerance of Cu/Zn superoxide dismutase and ascorbate peroxidase expressing transgenic tobacco seeds and seedlings against multiple abiotic stresses. *Int. J. Agric. Biol.* 15, 725-730.

Li, X., Gao, M.-J., Pan, H.-Yu, Cui, De-J., Gruber, M.Y., (2010). Purple canola: *Arabidopsis PAP1* increases antioxidants and phenolics in *Brassica napus* leaves. J. Agric. Food Chem. 58, 1639–1645.

Liu, H., Jiang, W., Bi, Y., Luo, Y., (2005). Postharvest BTH treatment induces resistance of peach (*Prunus persica* L. cv. Jiubao) fruit to infection by *Penicillium expansum* and enhances activity of fruit defense mechanisms. *Postharvest Biol. Technol.* 35, 263–269.

Luo, X., Wu, J., Li, Y., Nan, Z., Guo, X., Wang, Y., Zhang, A., Xia, G., Tian, Y. (2013). Synergistic effects of *GhSOD1* and *GhCAT1* overexpression in cotton chloroplasts on enhancing tolerance to methyl viologen and salt stresses. *PLoS ONE.* | www.plosone.org. | 8, e54002.

Mäkelä, P., Kärkkäinen, J., Somersalo, S., (2000). Effect of glycinebetaine on chloroplast ultrastructure, chlorophyll and protein content, and RuBPCO activities in tomato grown under drought or salinity. *Biol. Plantarum.* 43, 471-475.

Marino, D., Dunand, C., Puppo, A., Pauly, N., (2012). A burst of plant NADPH oxidases. Trends Plant Sci;17:9–15.

Martins, L.L., Mourato, M.P., Cardoso, A.I., Pinto, A.P., Mota, A.M., Gonçalves, M.L.S., de Varennes, A., (2011). Oxidative stress induced by cadmium in *Nicotiana tabacum* L.: effects on growth parameters, oxidative damage and antioxidant responses in different plant parts. *Acta Physiol. Plant.* 33, 1375–1383.

Matamoros, M.A., Loscos, J., Dietz, K.-J., Aparicio-Tejo, P.M., Becana, M., (2010). Function of antioxidant enzymes and metabolites during maturation of pea fruits. *J. Exp. Bot.* 61, 87-97.

Mittlera, R., Blumwald, E., (2015). The roles of ROS and ABA in systemic acquired acclimation. *The Plant Cell.* 27, 64-70.

Molina-Rueda, J.J., Tsai, C.J., Kirby, E.G., (2013). The Populus superoxide dismutase gene family and its responses to drought stress in transgenic poplar overexpressing a pine cytosolic glutamine synthetase (GS1a). PLOS ONE | www.plosone.org. Volume 8 | Issue 2 | e56421.

Murashige, T, Skoog, F., (1962). A revised medium for rapid growth and bioassays with tobacco tissue cultures. *Physiol Plant.* 15: 473-497.

Öncel, I., Yurdakulol, E., Keleş, Y., Kurt, L., Yıldız, A., (2004). Role of antioxidant defense system and biochemical adaptation on stress tolerance of high mountain and steppe plants. *Acta Oecologica.* 26, 211–218.

Pandey, P., Srivastava, R.K., Dubey, R.S. (2014). Water deficit and aluminum tolerance are associated with a high antioxidative enzyme capacity in Indica rice seedlings. *Protoplasma*. 251,147-60.

Pilon, M., Ravet, K., Tapken, W., (2011). The biogenesis and physiological function of chloroplast superoxide dismutases. *Biochim. Biophys. Acta*. 1807, 989-998.

Qiao, G., Zhou, J., Jiang, J., Sun, Y., Pan, L., Song, H., Jiang, J., Zhuo, R., Wang, X., Sun, Z., (2010). Transformation of *Liquidambar formosana* L. via *Agrobacterium tumefaciens* using a mannose selection system and recovery of salt tolerant lines. *Plant Cell Tiss Organ Cult.* 102, 163–170.

Randhir, R., Shetty, K., (2004). Microwave-induced stimulation of l-DOPA, phenolics and antioxidant activity in fava bean (*Vicia faba*) for Parkinson's diet. *Process Biochem*. 39, 1775–1784.

Reguera, M., Peleg, Z., Blumwald, E., (2012). Targeting metabolic pathways for genetic engineering abiotic stress-tolerance in crops. *Biochim. Biophys. Acta.* 1819, 186–194.

Sakhno, L.A., Gocheva, E.A., Komarnitskii, I.K., Kuchuk, N.V., (2008). Stable expression of the promoterless *bar* gene in transformed rapeseed plants. *Cytol Genet.* 42, 21–28.

Sakhno, L.O., (2015). Adaptive plasticity in osmotic stress of biotech canola (*Brassica napus* L.) possessing *cyp11A1* or simultaneously *desC* and *epsps* transgenes. *Naukovi dopovidi NUBIP*. 54. Available from http://nd.nubip.edu.ua/2015_5/9.pdf.

Sakhno, L.O., Komarnitsskii, I.K., Kuchuk, M.V., (2015). Inheritance of glyphosate and glufosinate resistance in T_1-T_2 generations of biotechnological canola (*Brassica napus*) plants. *The Bulletin of Vavilov Society of Geneticists and Breeders of Ukraine*. 13, 3-10.

Sakhno, L.O., Slyvets, M.S., (2014). Superoxide dismutase activity in transgenic canola. *Cytol Genet.* 48, 145-149.

Sakhno, L.O., Slyvets, M.S., Kuchuk, M.V., (2014). *Cyp*11A1 canola plants under heat stress conditions. *Cytol Genet.* 48, 279-284.

Sakhno, L.O., Morgun, B.V., Kvasko, O.Y., Kuchuk, M.V., (2010). Transformed canola plants expressing mammalian *cyp11A1* gene of cytochrome $P450_{SCC}$. *Biotechnol. Acta.* 3, 74–82.

Semenov, V.L., Yarosh, A.M., (1985). A method for determination of antioxidation activity of the biological material. *Ukr. Biochem. J.* 57, 50–52.

Slyvets, M., Sakhno, L., (2014). Human interferon alpha 2b positively affects canola growth in both aseptic non-stressed and water deficit conditions. *IJBSANS.* 1, 104-118.

Slyvets, M.S., Sakhno, L.O., Sheludko, Y.V., (2013). Expression of cyanobacterium *Des*C transgene does not change canola plant cold resistance. *The Bulletin of Vavilov Society of Geneticists and Breeders of Ukraine.* 11, 267-275.

Spivak, S.G., Berdichevets, I.N., Yarmolinsky, D. G., Maneshina, T. V., Shpakovski, G. V., Kartel, N. A. (2009). Construction and characteristics of transgenic tobacco *Nicotiana tabacum* L. plants expressing *CYP11A1* cDNA encoding cytochrome P450SCC. *Rus. J. Genetics*. 45, 1217-1224.

Stevenson, D.E., Lowe, T., (2009). Plant-derived compounds as antioxidants for health – are they all really antioxidants? *Funct Plant Sci Biotechnol.* 3 (Special Issue 1), 1-12.

Su, X., Zhou, P., Wang, R. Luo, Z., Xia, Z., (2015). Overexpression of the maize *psbA* gene enhances sulfur dioxide tolerance in transgenic tobacco. *Plant Cell Tiss Organ Cult.* 120, 303-311.

Sudipta, K. M., Swamy M.K., Balasubramanya, S., Anuradha, M., (2014). Assessment of genetic fidelity, antioxidant enzyme activity and proline content of micropropagated and field grown plants of *Leptadenia reticulate* (Wight&Arn.) – an endangered medicinal plant. *Plant Cell Biotech Molec Biol.* 15, 127–135.

Susuki, N., Miller, G., Morales, J., Shulaev, V., Torres, M.A., Mittler, R., (2011). Respiratory burst oxidases: the engines of ROS signaling. *Curr Opin Plant Biol.*14, 691–699.

Thornburg, R.W., Carter, C., Powell, A., Mittler, R., Rizhsky, L., Horner, H.T., (2003). A major function of the tobacco floral nectary is defense against microbial attack. *Plant Syst. Evol.* 238, 211–218.

Valko, M., Leibfritz, D., Moncol, J., Cronin, M.T.D., Mazur, M., Telser, J., (2007). Free radicals and antioxidants in normal physiological functions and human disease. *Int. J. Biochem. Cell B.* 39, 44–84.

Vencill, W.K., Nichols, R.L., Webster, T.M., Soteres, J.K., Mallory-Smith, C., Burgos, N.R., Johnson, W.G., McClelland, M.R., (2012). Herbicide resistance: toward an understanding of resistance development and the impact of herbicide-resistant crops. *Weed Sci.* Special Issue, 2–30.

What is Triazine Tolerant Roundup Ready canola? Available from http://roundupreadycanola.com.au/wp-content/uploads/2014/03/What-is-Triazine-Tolerant-Roundup-Ready-canola1.pdf.

Wojtera-Kwiczor, J., Zukowska, W., Graj, W., Małecka, A., Piechalak, A., Ciszewska, L., Chrzanowski, Ł., Lisiecki, P., Komorowicz, I., Barałkiewicz, D., Voss, I., Scheibe, R., Tomaszewska, B., (2014). Rhizoremediation of diesel-contaminated soil with two rapeseed varieties and petroleum degraders reveals different responses of the plant defense mechanisms. *Int. J. Phytoremed.* 16, 770–789.

Xie, Z., Duan, L., Tian, X., Wang, B., Eneji, A.E., Li, Z., (2008). Coronatine alleviates salinity stress in cotton by improving the antioxidative defense system and radical-scavenging activity. *J Plant Physiol.* 165, 375–384.

Xu, W., Guo, F., Zhou, X., Shang, Y., Yuan, Y., Zhang, F., Huang, K., (2011). Unintended effects were investigated in antioxidant activity between genetically modified organisms and their nontransgenic control. *Afr. J. Biotechnol.* 10, 9272-9279.

Yesilbag, D., Eren, M., Agel, H., Kovanlikaya, A., Balci, F., (2011). Effects of dietary rosemary, rosemary volatile oil and vitamin E on broiler performance, meat quality and serum SOD activity. *Brit. Poultry Sci.* 52, 472-482.

Reviewed by Anatoli Giritch, Nomad Bioscience GmbH, gyrych@nomadbioscience.com.

INDEX

A

B

C

D

G

H

I

J

K

L

M

N

O

P

Q

R

S

T

U

V

W

Y